STRUCTURALISM

STRUCTURALISM

Jean Piaget

Translated and edited by CHANINAH MASCHLER

Basic Books, Inc.

PUBLISHERS

NEW YORK

TRANSLATOR'S NOTE

I SHOULD LIKE TO THANK Professor Peter Caws of the City University of New York for reading and mending this translation. Against his, possibly better, judgment, I have encumbered the text with many footnotes and references which were not in Piaget's original text. My feeling was that without these the translation would fail in its primary purpose, that of orienting the uninitiated American or English reader.

The translation is at times more than free, but clarity and readability seemed to require this.

New York   CHANINAH MASCHLER

CONTENTS

STRUCTURALISM

I

INTRODUCTION AND LOCATION OF PROBLEMS

1. *Definitions*

Structuralism is often said to be hard to define because it has taken too many different forms for a common denominator to be in evidence: the structures invoked by the several "structuralists" have acquired increasingly diverse significations. Nevertheless, upon examining and comparing the various meanings it has acquired in the sciences and, unfortunately, at cocktail parties, a synthesis seems feasible, though only on the condition of separating two problems which, while always conjoined in fact, are logically distinct. The first of these problems is to make out the nature of the affirmative ideal that goes with the very idea of structure, the ideal manifested in the conquests and hopes of every variety of structuralism. The second is to describe and analyze the critical intentions attendant on the birth and development of any particular variety of structuralism.

To dissociate the two problems is to admit in effect that there really is an ideal of intelligibility held in common, or at least aspired after, by all structuralists, even though their critical objectives vary enormously. For the mathematicians, structuralism is opposed to compartmentalization, which it counteracts by recovering unity through isomorphisms. For several generations of linguists, structuralism is chiefly a departure from the diachronic study of isolated linguistic phenomena which prevailed in the nineteenth century and a turn to the investigation of synchronously functioning unified language systems.[1] In psychology, structuralism has long combatted the atomistic tendency to reduce wholes to their prior elements. And in current philosophical discussions we find structuralism tackling historicism, functionalism, sometimes even all theories that have recourse to the human subject.

So, obviously, if we try to define structuralism negatively, in terms of its opposition to other positions, and refuse to consider it apart from these, we shall find nothing but diversity and all the contradictions that are linked with the vagaries in the history of the sciences and of ideas. On the other hand, once we focus on the positive content of the idea of structure, we come upon at least two aspects that are common to all varieties of structuralism: first, an ideal (perhaps a hope) of intrinsic intelligibility supported by the postulate that structures

[1] For a definition of the terms "diachronic" and "synchronic" see the quotation from Saussure in Note 1, Chapter V. [Trans.]

are self-sufficient and that, to grasp them, we do not have to make reference to all sorts of extraneous elements; second, certain insights—to the extent that one has succeeded in actually making out certain structures, their theoretical employment has shown that structures in general have, despite their diversity, certain common and perhaps necessary properties.

As a first approximation, we may say that a structure is a system of transformations. Inasmuch as it is a system and not a mere collection of elements and their properties, these transformations involve laws: the structure is preserved or enriched by the interplay of its transformation laws, which never yield results external to the system nor employ elements that are external to it. In short, the notion of structure is comprised of three key ideas: the idea of wholeness, the idea of transformation, and the idea of self-regulation.

The discovery of structure may, either immediately or at a much later stage, give rise to formalization. Such formalization is, however, always the creature of the theoretician, whereas structure itself exists apart from him. Formalization sometimes proceeds by direct translation into logical or mathematical equations, sometimes passes through the intermediate stage of constructing a cybernetic model, the level of formalization depending upon the choice of the theoretician. But, it is worth repeating, the mode of existence of the structure he earlier discovered must be determined separately for each particular area of investigation.

The notion of transformation allows us to delimit the problem of definition in a preliminary way; for if it were necessary to cover formalism in every sense of the word by the idea of structure, all philosophical positions that are not strictly empiricist would be let in again—those which invoke Platonic forms or Husserlian essences, not to forget Kant's brand of formalism, and even several varieties of empiricism (for example, the logical positivism of the Vienna Circle, who stress syntactic and semantic forms in their analysis of logic). Now in the narrower sense we are about to define, current logical theory only rarely takes account of "structures," for in many ways it has remained subservient to a rather stubborn atomistic tendency and is only beginning to open up to structuralism.

In this little book we shall, therefore, confine ourselves to the kinds of structuralism that are to be met in mathematics and the several empirical sciences, already a sufficiently venturesome undertaking. And in conclusion we shall take up some of the philosophical movements more or less inspired by the various kinds of structuralism that have sprung up in the social sciences. But first we must elaborate somewhat on the definition of structuralism in general that is here proposed, else it will be hard to understand why a notion as abstract as that of a "system closed under transformation" should raise such high hopes in all domains of inquiry.

2. *Wholeness*

That wholeness is a defining mark of structures almost goes without saying, since all structuralists—mathematicians, linguists, psychologists, or what have you—are at one in recognizing as fundamental the contrast between *structures* and *aggregates,* the former being wholes, the latter composites formed of elements that are independent of the complexes into which they enter. To insist on this distinction is not to deny that structures have elements, but the elements of a structure are subordinated to laws, and it is in terms of these laws that the structure *qua* whole or system is defined. Moreover, the laws governing a structure's composition are not reducible to cumulative one-by-one association of its elements: they confer on the whole as such over-all properties distinct from the properties of its elements. Here is an example of what we have in mind: the integers do not exist in isolation from one another, nor were they discovered one by one in some accidental sequence and then, finally, united into a whole. They do not come upon the scene except as ordered, and this order of the integers is associated with *structural* properties (of groups, fields, rings, and the like), which are quite different from the properties of number individuals, each of which is even or odd, prime or non-prime, and so on.

The idea of wholeness does, however, raise a good many problems, of which we shall take up just the two principal ones, the first bearing on its *nature,* the other on its mode of *formation* (or preformation).

It would be a mistake to think that, in all domains, the epistemological alternatives reduce to just two options: either admit wholes defined in terms of their structural laws, or allow only for atomistic compounding of prior elements. No matter what area of science we subject to scrutiny, whether we consider the perceptual structures of the Gestalt psychologists or the social wholes (classes or entire societies) of sociologists and anthropologists, we find that not one but two alternatives to atomism have made their way in the history of ideas, only one of which appears to us in tune with the spirit of modern structuralism.

The first consists in simply reversing the sequence that appeared natural to those who wanted to proceed from the simple to the complex (from sense impressions to perceptual complexes, from individuals to social groups, and so forth). The whole which this sort of critic of atomism posits at the outset is viewed as the outcome of some sort of emergence, vaguely conceived as a law of nature and not further analyzed. Thus, when Comte proposed to explain men in terms of humanity, not humanity in terms of men, or when Durkheim thought of the social whole as emerging from the union of individuals in much the same way as molecules are formed by the union of atoms, or when the Gestalt psychologists believed they could discern immediate wholes in primary perception comparable to the field effects that figure in electromagnetism, they did indeed remind us that a whole is not the same as a simple juxtaposition of previously available elements, and for this they deserve our gratitude; but by viewing the whole as prior to its elements or contemporaneous with their "contact," they simplified the problem to such an extent as to risk bypassing all the central questions—questions about the nature of a whole's laws of composition.

Over and beyond the schemes of atomist association on the one hand and emergent totalities on the other, there is, however, a third, that of operational structuralism. It adopts from the start a relational perspective, according to which it is neither the elements nor a whole that comes about in a manner one knows not how, but the relations among elements that count. In other words, the logical procedures or natural processes by which the whole is formed are primary, not the whole, which is consequent on the system's laws of composition, or the elements.

But at this point a second, and much more serious, problem springs up, the really central problem of structuralism: Have these composite wholes always been composed? How can this be? Did not someone compound them? Or were they initially (and are they still) in *process* of composition? To put the question in a different way: Do structures call for *formation,* or is only some sort of eternal *preformation* compatible with them?

Structuralism, it seems, must choose between structure-less genesis on the one hand and ungenerated wholes or forms on the other; the former would make it revert to that atomistic association to which empiricism has accustomed us; the latter constantly threaten to make it lapse into a theory of Husserlian essences, Platonic forms, or Kantian *a priori* forms of synthesis. Unless, of course, there is a way of passing between the horns of this dilemma.

As is to be expected, it is on this problem that opinion is most divided, some going so far as to contend that the problem of the genesis of structures cannot so much as be formulated because structure is of its very nature non-temporal (as if this were not in its own way a solution of the problem, namely, the choice of a preformational view of the origin of structures).

Actually, the problem we now are discussing arises with the notion of wholeness itself. It can be narrowed down once we take the second characteristic of structures, namely, their being systems of transformations rather than of static forms, seriously.

3. *Transformations*

If the character of structured wholes depends on their laws of composition, these laws must of their very nature be *structuring:* it is the constant duality, or bipolarity, of always being simultaneously *structuring* and *structured* that accounts for the success of the notion of law or rule employed by structuralists. Like Cournot's "order"[2] (a special case of the structures treated in modern algebra), a structure's laws of composition are defined "implicitly,"[3] i.e., as governing the transformations of the system which they structure.

When one considers the history of structuralism in linguistics and psychology, this last statement may seem somewhat surprising: In linguistics, structuralism started with the work of Saussure, which does not seem to bear out our claim; moreover, Saussure used the single word "system" to cover both laws of synchronic opposition

[2] The reference is to Augustin Cournot's *Researches into the Mathematical Principles of Wealth* (1838; reprinted, New York: Kelley, 1927), the first systematic treatise on mathematical economics. See Oskar Morgenstern's article on mathematical economics in *Encyclopaedia of the Social Sciences,* V, 364ff. [Trans.]

[3] For a brief account of the notion of "implicit definition" see, for example, Hermann Weyl, *Philosophy of Mathematics and Natural Science* (Princeton: Princeton University Press, 1949), pp. 24ff. [Trans.]

and laws of synchronic equilibration.[4] In Gestalt psychology, the perceptual forms that are said to have a Gestalt character are, generally speaking, static. But it is unwise to view an intellectual current exclusively in terms of its origin; it should be seen in its flow. Besides, even in their beginnings, linguistic and psychological structuralism were associated with the dawning of ideas of transformation. Synchronic language systems are not immobile; such systems exclude or allow for novelty (acceptance or exclusion being a function of requirements determined by the system's laws of opposition and connection). And it did not take long before Saussure's notion of some sort of dynamic equilibrium became elaborated into a theory of style by Bally.[5] Admittedly, we are not yet in the presence of Chomsky's "transformational grammar," but Bally's stylistic is headed toward the idea of transformation, at least in the weak sense of "individual variation." As for psychological *Gestalten*, their inventors from the beginning spoke of laws of organization by which the sensory given is transformed, and the probabilistic interpretation we can today give of such laws accentuates this transformational aspect of perception.

Indeed, all known structures—from mathematical groups to kinship systems—are, without exception, systems of transformation. But transformation need not be a temporal process: $1 + 1$ "make" 2; 3 "follows hard on"

[4] See, for example, pp. 107, 117, 119–122 in Saussure's *Course in General Linguistics*, ed. C. Bally and A. Séchehaye, trans. W. Baskin (New York: Philosophical Library, 1959). [Trans.]

[5] C. Bally, *Précis de stylistique* (Geneva, 1905) and *Traité de stylistique française* (Heidelberg, 1909). [Trans.]

2; clearly, the "making" and "following" here meant are not temporal processes. On the other hand, transformation can be a temporal process: getting married "takes time." Were it not for the idea of transformation, structures would lose all explanatory import, since they would collapse into static forms.

The very centrality of the idea of transformation makes the question of origin, that is, of the relation between transformation and formation, inevitable. Certainly, the elements of a structure must be differentiated from the transformation laws which apply to them. Because it is the former which undergo transformation or change, it is easy to think of the latter as immutable. Even in varieties of structuralism which are not formalized in the strict sense, one finds outstanding workers who are so little concerned about psychological origins that they jump straight from the *stability* of transformation rules to their *innateness*. Noam Chomsky is a case in point: for him generative grammars appear to demand innate syntactic laws, as if stability could not be explained in terms of equilibrium mechanisms, and as if the appeal to biology implied by the hypothesis of innateness did not pose problems of formation just as complex as those involved in a psychological account.

Now the implicit hope of anti-historical or anti-genetic structuralist theories is that structure might in the end be given a non-temporal mathematical or logical foundation (Chomsky has, in fact, reduced his grammars to a formal "monoid" structure).[6] But if what is wanted is a

----

[6] The reference here, as also on pp. 71 and 78, may be to Chomsky's paper in the *Handbook of Mathematical Psychology*, ed. R. D. Luce, R. R. Bush, and E. Galanter (New York: John

general theory of structure, such as must meet the requirements of an interdisciplinary epistemology, then one can hardly avoid asking, when presented with such non-temporal systems as a mathematical group or a set of subsets, how these systems were *obtained*—unless, of course, one is willing to stay put in the heavens of transcendentalism. One could proceed by postulate, as in axiomatic systems, but from the epistemological point of view this is an elegant way of cheating which takes advantage of the prior labor of those who constructed the intuitive system without which there would be nothing to axiomatize. As we shall see in Chapter II, Gödel's conception of a structure's relative "power" or "weakness" introduces a genealogical relation among structures which provides a method less open to epistemological objections. But once we take this tack the central problem, not as yet of the history or psychogenesis of structures but of their *construction,* and of the relation between *structuralism* and *constructivism,* is no longer avoidable. This will, then, be one among our several themes.

4. *Self-Regulation*

The third basic property of structures is, as we said, that they are self-regulating, self-regulation entailing self-

Wiley, 1963–1965), where he says (2.274): "A set that includes an identity and is closed under an associative law of composition is called a *monoid.* Because monoids satisfy three of the four postulates of a group, they are sometimes called *semigroups.* A *group* is a monoid, all of whose elements have inverses." *Syntactic Structures* makes no mention of "monoids." [Trans.]

maintenance and closure. Let us start by considering the two derivative properties: what they add up to is that the transformations inherent in a structure never lead beyond the system but always engender elements that belong to it and preserve its laws. Again an example will help to clarify: In adding or subtracting any two whole numbers, another whole number is obtained, and one which satisfies the laws of the "additive group" of whole numbers. It is in this sense that a structure is "closed," a notion perfectly compatible with the structure's being considered a substructure of a larger one; but in being treated as a substructure, a structure does not lose its own boundaries; the larger structure does not "annex" the substructure; if anything, we have a confederation, so that the laws of the substructure are not altered but conserved and the intervening change is an enrichment rather than an impoverishment.

These properties of conservation along with stability of boundaries despite the construction of indefinitely many new elements presuppose that structures are self-regulating. There can be no doubt that it is this latter conception which makes the idea of structure so important and which accounts for the high hopes it raises in all domains of inquiry: Once an area of knowledge has been reduced to a self-regulating system or "structure," the feeling that one has at last come upon its innermost source of movement is hardly avoidable. Now self-regulation may be achieved by various procedures or processes, and these can be ranked in order of increasing complexity; we are thus brought back to our earlier question about a system's construction, i.e., in the last analysis, its formation.

At the highest level (though it should be remembered that what we call the top of the pyramid may be viewed as its base by others) self-regulation proceeds by the application of perfectly explicit rules, these rules being, of course, the very ones that define the structure under consideration. It might therefore be urged against us that talk about self-regulation is quite empty, since it refers either to the laws of the structure under consideration, and it goes without saying that they "regulate" it, or it refers to the mathematician or logician who "operates" on the elements of the system, and that he operates correctly under normal circumstances again goes without saying. Granted all this, there still remains the question: Just what *is* an operation structurally considered? From the cybernetic point of view, an operation is a "perfect" regulation. What this means is that an operational system is one which excludes errors before they are made, because every operation has its inverse in the system (e.g., subtraction is the inverse of addition, $+n - n = 0$) or, to put it differently, because every operation is reversible, an "erroneous result" is simply not an element of the system (if $+n - n \neq 0$, then $n \neq n$).

But there is, of course, an immense class of structures which are not strictly logical or mathematical, that is, whose transformations unfold in time: linguistic structures, sociological structures, psychological structures, and so on. Such transformations are governed by laws ("regulations" in the cybernetic sense of the word) which are not in the strict sense "operations," because they are not entirely reversible (in the sense in which multiplication is reversible by division or addition by subtraction). Transformation laws of this kind depend upon the interplay of anticipation and correction (feedback). As we shall see in Section 10, the range of application of feedback mechanisms is enormous.

Finally, there are regularities in the non-technical sense of the word which depend upon far simpler structural mechanisms, on rhythmic mechanisms such as pervade biology and human life at every level.[7] Rhythm too is self-regulating, by virtue of symmetries and repetitions. Though the self-regulation that is here involved is of a much more elementary sort, it would not do to exclude rhythmic systems from the domain of structure.

Rhythm, regulation, operation—these are the three basic mechanisms of self-regulation and self-maintenance. One may, if one so desires, view them as the "real" stages of a structure's "construction," or, reversing the sequence, one may use operational mechanisms of a quasi-Platonic and non-temporal sort as a "basis" from which the others are then in some manner "derived."

---

[7] The study of such biological rhythms and periodicities (i.e., cycles of approximately 24 hours, which are remarkably general) has in recent years been turned into an entire new discipline with its own specialized mathematical and experimental techniques.

II

MATHEMATICAL AND LOGICAL STRUCTURES

5. *Groups*

A critical account of structuralism must begin with a consideration of mathematical structures, not only for logical but even for historical reasons. True, when structuralism first made its appearance in linguistics and psychology, the formative influences were not directly mathematical—Saussure's concept of synchronic equilibrium was inspired by ideas then current in economic theory, and the Gestalt psychologists took off from physics. But the structural models of Lévi-Strauss, the acknowledged master of present-day social and cultural anthropology, are a direct adaptation of general algebra. Moreover, if we accept the general definition of structure sketched in the preceding chapter, the first known "structure," and the first to be studied as such, was surely the mathematical "group," Galois' discovery, which little by little conquered the whole of nineteenth-century mathematics.[1]

A mathematical group is a system consisting of a set of elements (e.g., the integers, positive and negative) together with an operation or rule of combination (e.g., addition) and having the following properties:

1. performed upon elements of the set, the combinatory operation yields only elements of the set;

2. the set contains a neuter or identity element (in the given case, 0) such that, when it is combined with any other element of the set, the latter is unaffected by the combinatory operation (in the given case, $n + 0 = n$ and, since addition is commutative, $n + 0 = 0 + n = n$);

3. the combinatory operation has an inverse in the system (here subtraction) such that, in combination with the former, the latter yields the neuter or identity element $(+n - n = 0)$;

4. the combinatory operation (and its inverse) is associative $([n + m] + 1 = n + [m + 1])$.

Groups are today the foundation of algebra. The range and fruitfulness of the notion are extraordinary. We run into it in practically every area of mathematics and logic. It is already being used in an important way in physics, and very likely the day will come when it acquires a

---

[1] For an informal account of group theory, the reader might turn to Part IX, "The Supreme Art of Abstraction: Group Theory," in volume III of James R. Newman's *World of Mathematics* (New York: Simon and Schuster, 1956); for a somewhat more formal treatment, see Raymond L. Wilder, *Introduction to the Foundations of Mathematics* (New York: John Wiley, 1960), Chapter VII. [Trans.]

central role in biology as well. Clearly, then, we should try to understand the reasons for the immense success of the group concept. Since groups may be viewed as a kind of prototype of structures in general, and since they are defined and used in a domain where every assertion is subject to demonstration, we must look to them to ground our hope for the future of structuralism.

The primary reason for the success of the group concept is the peculiar—mathematical or logical—form of abstraction by which it is obtained; an account of its formation goes far to explain the group concept's wide range of applicability. When a property is arrived at by abstraction in the ordinary sense of the word, "drawn out" from things which have the property, it does, of course, tell us something about these things, but the more general the property, the thinner and less useful it usually is. Now the group concept or property is obtained, not by this sort of abstraction, but by a mode of thought characteristic of modern mathematics and logic—"reflective abstraction"—which does not derive properties from *things* but from our ways of *acting on things,* the operations we perform on them; perhaps, rather, from the various fundamental ways of *coordinating* such acts or operations—"uniting," "ordering," "placing in one-one correspondence," and so on. Thus, when we analyze the concept of groups, we come upon the following very general coordinations among operations:

1. the condition that a "return to the starting point" always be possible (via the "inverse operation");

2. the condition that the same "goal" or "terminus" be attainable by alternate routes and without the itin-

erary's affecting the point of arrival (associativity").[2]

Because of these two restrictive conditions, group structure makes for a certain coherence—whatever has that structure is governed by an internal logic, is a self-regulating system. This self-regulation is really the continual application of three of the basic principles of rationalism: the principle of non-contradiction, which is incarnate in the reversibility of transformations; the principle of identity, which is guaranteed by the permanence of the identity element; and the principle, less frequently cited but just as fundamental, according to which the end result is independent of the route taken. To illustrate the last point, consider the set of displacements in space. It constitutes a group (since any two successive displacements yield a displacement, a given displacement can always be "annulled" by an inverse displacement or "return," etc.). That the associativity of the group of spatial displacements (equivalent to our intuitive notion of using a detour) is absolutely essential is seen as soon as it is recognized that, if termini did vary with the paths traversed to reach them, space would lose its coherence and thereby be annihilated; what we would have instead would be some sort of perpetual Heraclitean flux.

Group structure and transformation go together. But when we speak of transformation, we mean an intelligible change, which does not transform things beyond recognition at one stroke, and which always preserves invariance in certain respects. To return to our example, the

[2] If the group operation and its inverse are commutative, the group is commutative or "abelian"; otherwise it is non-commutative.

displacement of a solid in ordinary space leaves its dimensions unchanged; similarly, the sum of the parts of a whole remains invariant under continual division. The existence of groups reveals how contrived the antithesis of self-sameness and change, on which Meyerson based his entire epistemology, really is; whereas according to him identity alone is rational, all change irrational, the consistency of the group concept, which calls for a certain inseparable connection of identity and change, is proof of their compatibility.

It is because the group concept combines transformation and conservation that it has become the basic constructivist tool. Groups are systems of transformations; but more important, groups are so defined that transformation can, so to say, be administered in small doses, for any group can be divided into subgroups and the avenues of approach from any one to any other can be marked out. Thus, starting with the group of which we spoke just now, the group of displacements, which leaves not only the dimensions of the displaced body or figure invariant, but preserves its angles, parallels, straight lines, and so on, as well, we can go on to the next "higher" group by letting the dimensions vary while preserving the other properties enumerated. In this way we obtain the group of similar figures of bodies: shape is kept invariant under transformation of dimensions. The group of displacements has thereby become a subgroup of the shape group. Next we may allow the angles to vary while conserving parallels and straight lines. A still more general group, that treated by "affine geometry" (which deals with such problems as how to transform one lozenge into another), is thus obtained, of which the shape group now becomes a subgroup. Continuing this process, parallels may be modified while straight lines are preserved; the "projective" group is thereby constructed; and the entire preceding series now becomes a "stack" of subgroups within the projective group. Finally, even straight lines may be subjected to transformation. Shapes are now treated as if they were elastic: only "biunique" and "bicontinuous" correspondence among their points are preserved under transformation. The group thus obtained (that of "homeomorphs") is the most general. It constitutes the subject matter of topology. The various kinds of geometry[3]—once taken to be static, purely representational, and disconnected from one another—are thus reduced to one vast construction whose transformations under a graded series of conditions of invariance yield a "nest" of subgroups within subgroups. It is this radical change of the traditional representational geometry into one integrated system of transformations which constitutes Felix Klein's famous Erlanger Program. The Erlanger Program[4] is a prime example of the scientific fruitfulness of structuralism.

---

[3] The various metric geometries—Euclidean and non-Euclidean—can be constructed by applying a "general metrics" to topology; that is, it is possible to "reverse directions" and to descend from the group of maximum generality all the way down to the group of displacements with which our account started.

[4] See Felix Klein, *Elementary Mathematics from an Advanced Standpoint: Geometry,* trans. E. R. Hedrick and C. A. Noble (New York: Dover, 1939). [Trans.]

6. *"Parent Structures"*

But in the eyes of contemporary structuralist mathematicians, like the Bourbaki,[5] the Erlanger Program amounts to only a partial victory for structuralism, since they want to subordinate all mathematics, not just geometry, to the idea of structure.

Classical mathematics is a quite heterogeneous collection of algebra, theory of numbers, analysis, geometry, probability calculus, and so on. Each of these has its own delimited subject matter; that is, each is thought to deal with a certain "species" of objects. The mathematicians of the Bourbaki circle, having noted that sets of the most various sorts, not just algebraic sets, may display the group property, and intent upon overcoming the traditional compartmentalization of mathematics into areas that exist simply side by side, initiated a program of generalization whereby the group structure becomes only one among a variety of basic structures. If the term "element" is applicable to objects as abstract as numbers, displacements, projection, and so on (some of which are, as we have seen, resultants of operations as well as operators), this means that group structure is quite independent of the intrinsic nature of its elements, which can, accordingly, be left unspecified. Transformations may be disengaged from the objects subject to such transformation and the group defined solely in terms of the set of trans-

---

[5] "Nicolas Bourbaki" is the collective pseudonym of a group of French mathematicians who publish under that name. [Trans.]

formations. The Bourbaki program consists essentially in extending this procedure by subjecting mathematical elements of every variety, regardless of the standard mathematical domain to which they belong, to this sort of "reflective abstraction" so as to arrive at structures of maximum generality.

It is worth noting that in its initial stages execution of the Bourbaki program called for some sort of "induction," since neither the form of the basic structures sought nor their number was known *a priori*. This quasi-inductive procedure led to the discovery of three "parent structures," that is, three not further reducible "sources" of all other structures.[6] The first of these is the *algebraic* structure. The prototype of this family is the mathematical group, together with all its derivatives—rings, fields, and so on. The characteristic of structures belonging to the algebraic family is that "reversibility" takes the form of "inversion" or "negation" (if T is the operator and T^{-1} its inverse, $T \cdot T^{-1} = 0$ in all algebraic structures). Next there are *order* structures. The "lattice" or "network" is their prototype, a structure as general as the group structure, but one which was not studied until comparatively recently (by Birkhoff, Glivenko, and others). Networks unite their elements by the predecessor/successor relation, any two elements of the network having a smallest upper bound (the nearest of the successors, or *supremum*) and a greatest lower bound (the closest predecessor, or *infimum*). Like the group property, the lattice property has a very wide range of application: it

[6] The number 3 may cause suspicion, so let us repeat—that there are just three such principal structures was a discovery, the outcome of "regressive analysis," not a postulate.

applies, for example, to the set of subsets of a collection, or to a group of its subgroups.[7] The defining mark of order structures is that reversibility takes the form, not of inversion, as in groups, but of reciprocity: "$(A \cdot B)$ precedes $(A + B)$" transforms into "$(A + B)$ succeeds $(A \cdot B)$" by permutation of the "$+$" and "$\cdot$" operators and the predecessor and successor relations. Finally, there are the topological structures of which we spoke in the preceding section: neighborhood, continuity, and limit are here the basic conceptions.

Once these three parent structures have been distinguished and characterized, the rest follow, that is, can be constructed. There are two methods of construction, combination and differentiation: a set of elements may be subjected to the restrictive conditions of two parent structures at the same time (algebraic topology, for instance, is yielded by combining algebraic and topological conditions); or substructures of any one of the three parent structures may be defined by introducing certain additional restrictive conditions. (This is the procedure we saw at work in Section 5 above.) The reverse procedure, "de-differentiation," is also possible: one may drop a restrictive condition and thereby move from a "stronger" to a "weaker" structure: the semigroup, for instance, may be defined as the structure resulting from the deletion of conditions 2 and 4 in the group definition given on page 18 above; the natural numbers greater than 0 constitute such a semigroup.

[7] A set S of n elements has 2^n subsets, since the set of subsets is obtained by taking the elements one by one, two by two, and so on, and the null set as well as the set S itself are counted as subsets.

Does this "mathematical architecture"[8] build on foundations that are in some manner "natural," or are the Bourbaki parent structures simply an axiomatic basis of their system? The question is not only interesting in its own right but will help us tie things together and clarify the general significance of structures. If we take the adjective in roughly the sense in which we speak of the positive whole numbers as "natural," the Bourbaki parent structures do appear to be natural. The positive whole numbers antedate mathematics; they are constructed by means of operations that stem from ordinary, everyday activities such as the "matching" to which even very primitive societies resort in their barter transactions, or at which we catch the playing child. When Cantor defined the first cardinal infinite in terms of one-one correspondence, he utilized an operation which, in its "natural" form, precedes nineteenth-century mathematics by uncounted millennia. Now when we study the intellectual development of the child, we find that the earliest cognitive operations, those which grow directly out of handling things, can be divided into precisely three large categories, according to whether reversibility takes the form of "inversion," of "reciprocity," or of "continuity" and "separation." Corresponding to the first—formally considered, algebraic structures—there are classificatory and number structures; corresponding to the second—formally considered, order structures—there are series and serial correspondences; corresponding to the last—formally considered, topological structures—there are operations that yield classes, not in terms of resemblances

[8] We borrow the expression from Bourbaki literature.

and differences, but in terms of "neighborhoods," "continuity," and "boundaries." It is remarkable that, psychogenetically, topological structures antedate metric and projective structures, that psychogenesis inverts the historical development of geometry but matches the Bourbaki "genealogy"!

These facts seem to suggest that the mother structures of the Bourbaki correspond to coordinations that are necessary to all intellectual activity, though they be very elementary, even rudimentary, and quite lacking in generality in the earliest stages of intellectual development. It would, in fact, not be difficult to show that in these very early stages intellectual operations grow directly out of sensory-motor coordinations, and that *intentional* sensory-motor acts—the human baby's or the chimpanzee's—*cannot* be understood apart from "structures" (see Chapter IV).

Before we sketch the implications of the foregoing observations for logic, we want to call the reader's attention to the fact that the structuralism of the Bourbaki circle is in process of transformation, under the influence of a current of thought well worth noting because it shows how new structures are, if not "formed," at least "discovered." What we have in mind are the "categories" of MacLane, Eilenberg, and others. The "categories" of the new branch of the Bourbaki school are classes which comprise "functions" and therefore "morphisms" among their elements (a function, in the usual acceptation of the word, being the "application" of one set to another [or itself], obviously engenders isomorphisms, in fact, every variety of "morphism"). Suffice it to say here that the "categories," with their emphasis upon functions, no longer revolve around parent structures but around the acts of correlation by which the latter were obtained: these new structures are not built up by in some way compounding the beings yielded by certain antecedent operations but by correlating these very operations, the formational procedures themselves.

There is some justice, then, in S. Papert's observation that MacLane's categories are a device for laying hold of mathematical operations rather than of "mathematics itself": they constitute yet another example of that reflective abstraction which derives its substance, not from objects, but from operations performed upon objects, even when the latter are themselves products of reflective abstraction.

These facts have an important bearing both on the nature and on the manner of construction of structures.

7. *Logical Structures*

Since logic is concerned with the form of knowledge, not its content, it is *prima facie* a privileged domain for structures. Yet, as we hinted earlier, this is not so if by "logic" we mean "mathematical" or "symbolic" logic, the only logic that really counts today. On the other hand, if we start from a rather broader perspective, such as allows us to raise the problem of "natural logic" (in approximately the sense in which we spoke of "natural numbers" in Section 6), we find that the "opposites" in terms of which we just now characterized logic, namely, "form" and "content," are correlatives, not absolutes: the "contents" on which logical forms are imposed are not formless; they have forms of their own; else they could not "potentially be logicized." And the forms of what originally appeared to *be* "pure content" in turn themselves *have* content, though less distinctly made out, a content with its own form, and so on, indefinitely, each element being "content" relative to some prior element and "form" for some posterior element.

For structuralist theory this "nesting" of forms and the correlativity of form and content are of the utmost significance, but from the logical point of view these facts are of no interest, unless indirectly, through their bearing on the problem of the limits of formalization.

Symbolic logic proceeds as follows: it assumes some arbitrary position in the ascending contents/form series and turns its starting point into an absolute beginning—into the "basis" of a "logical system." More explicitly, the basis of such an axiomatic system consists of (1) certain primitive or undefined conceptions, which serve to define the rest; (2) certain axioms or undemonstrated propositions, which serve to demonstrate the rest. The undefined conceptions are primitive or indefin*able* and the undemonstrated propositions axiomatic or indemonstr*able* within the particular system under consideration, but they may well turn up derived in some other system. And the axiomatic method leaves the logician free to choose what system he pleases; all it requires is that the primitive conceptions and the propositions which serve as axioms be "adequate," compatible, and "mutually independent" (that is, reduced to a minimum). Next, there will have to be certain rules of construction, i.e., formation and transformation procedures. The formal system thus obtained is self-sufficient, dispenses with intuition, and is, in the sense explained, "absolute." There remains, of course, the problem of the limits of formalization as well as the epistemological question of how the primitive conceptions and the axioms were obtained, but from the formal point of view, which is that of the logician, such an axiomatic system is a perfect and no doubt unique example of "autonomy"—the rules are wholly "internal," and all regulation is self-regulation.

It might, accordingly, be maintained that any such system (they are numberless) is a "structure," since it meets the requirements of the definition offered on page 5. But from the structuralist perspective the logician's formal systems are wanting in at least two respects. In the first place, they are fabricated *ad hoc,* and, whether this be openly acknowledged or not, what structuralism is really after is to discover "natural structures"—some using this somewhat vague and often denigrated word to refer to an ultimate rootedness in human nature, others, on the contrary, to indicate a non-human absolute to which we must accommodate ourselves instead of the reverse. (The former obviously risk a lapse into a priorism, the latter a return to transcendent essences.) But there is a more serious problem: a logical system, though a closed whole with respect to the theorems it demonstrates, is nevertheless only a relative whole; it remains "open" with respect to those formulae which, though recognized as true when one goes "up" to its metatheory, are nevertheless indemonstrable so long as one stays "in" the system; and, since the primitive conceptions and axioms have all sorts of implicit elements, the system is "open" at the "bottom" as well.

"Logical structuralism," if we may so call it, has chiefly

been concerned with this latter problem, its announced objective being the recovery of what lies "beneath" the operations codified by axioms. Full-fledged structures not just analogous to the major intuitive structures employed by mathematicians but identical with some of these turn out to furnish the underpinnings of logic, so that logic becomes a part of the theory of structure which is today called general algebra.[9] What makes these struc-

[9] Boole's "Investigations of the Laws of Thought" (1853) which, as is well known, takes the form of an algebra, to this day called "Boolean algebra," furnishes a particularly interesting example of this structural identity of logic and algebra. The Boolean algebra can be interpreted as a logic of classes, a propositional logic, or a two-valued arithmetic. By adding certain restrictive conditions—the law of distribution, complementarity, and the existence of a maximum and a minimum element—to ordinary networks, Boolean algebra may also be viewed as yielding a special kind of network, the Boolean. Besides, the two Boolean operations, that is [for propositional calculus] exclusive disjunction (p or q but not both) and equivalence (p and q or neither p nor q) both yield a group, either one of which can be transformed into a commutative ring (see J. B. Grize, "Logique," in *Logique et connaissance scientifique*, XXII, 277, of *Encyclopédie de la Pléiade* by Piaget *et al.*). Thus, two of the principal structures of modern algebra turn up again in logic. But a group of still greater generality, namely a particular variety of Klein's quaternaries, can be discerned in Boolean algebra, the *INRC* group we described in our *Traité de Logique* (Colin, 1949). If we start with an operation like the implication $p \supset q$, we obtain its negation $p \cdot \bar{q}$, by inverting it, N; by permutation of its terms (or simply by preserving the conditional's form but replacing p by $\bar{p}$ and q by $\bar{q}$) we obtain $q \supset p$, its reciprocal, R; if, starting with $p \supset q$ in its "normal form" ($p \cdot q \vee \bar{p} \cdot q \vee \bar{p} \cdot \bar{q}$), we interchange "$\vee$" and "$\cdot$" we obtain $\bar{p} \cdot q$., the correlative of $p \supset q$, C; finally, if we leave $p \supset p$ untouched, we get the identity transformation, I. Thus, commutatively, $NR = C$; $NC = R$; $CR = N$; $NRC = I$. What we have here is a group of four transformations of which the operations of a two-valued propo-

tures all the more interesting is that their psychological development can be traced by studying the growth of natural thought. But this brings up problems which we must reserve for a later chapter.

8. *The Limits of Formalization*

Reflection upon logical structures provides us with an opportunity to see how structures differ from their formalized counterparts and proceed from a "natural" reality. But to explain ourselves here we must go slowly.

In 1931 Kurt Gödel made a discovery which created a tremendous stir, because it undermined the then prevailing formalism,[10] according to which mathematics was reducible to logic and logic could be exhaustively formal-

---

sitional logic supply as many instances as one can form quaternaries from the elements of its set of subsets. For some of these, $I = R$ and $N = C$ or $I = C$ and $N = R$, but of course $I = C$ is always false.

Marc Barbut's commentary on our earlier description of the INRC group ("Problèmes du structuralisme," *Les Temps modernes*, No. 246 (November 1966), p. 804) may give rise to a misunderstanding: The quaternary INRC should not be assimilated to the simpler form where for AB the three other transformations are reducible to (1) changing A, (2) changing B, (3) changing A and B both. The group INRC has for its elements, not the 4 cases of a truth table for 2 variables, but the 16 combinations of its set of subsets (or, for 3 variables, the 256 combinations, and so on). Because of its greater complexity, the INRC group does not make an appearance psychologically until early adolescence, whereas Barbut's simpler models of groups of 4 elements are accessible to 7 and 8 year olds.

[10] See Ernest Nagel, "The Formation of Modern Conceptions of Formal Logic in the Development of Geometry," *Osiris*, VII, 149ff. [Trans.]

ized. Gödel established definitively that the formalist program cannot be executed. In the first place, he showed that no consistent formal system sufficiently "rich" to contain elementary arithmetic (for example, the system of Russell and Whitehead's *Principia Mathematica*), can, by its own principles of reasoning (*a fortiori,* by those of relatively "weaker" systems) *demonstrate* its own consistency; second, that any such system allows for the generation of propositions which are "formally undecidable," or, to use yet another technical expression, that any logical system that might appear capable of serving as foundation for mathematics is "essentially incomplete."[11] Though it was later discovered, by Gentzen, that consistency proofs of elementary arithmetic can be furnished by employing principles of reasoning "stronger" than those used within arithmetic, the consistency of these stronger rules of inference—roughly, those of Cantor's transfinite arithmetic—can only be demonstrated by appealing to a logical theory of yet a higher rank. In other words, since Gödel we know that the axiomatic method has certain inherent limitations, though these limits can be "shifted" by shifting systems.

The first noteworthy effect of these developments was that a notion of greater or lesser "power" entered into the domain of structures, into that delimited subarea where structures can actually be compared in terms of their relative "strength" or "weakness." The hierarchy that was

[11] For a non-technical but accurate account of Gödel's paper "On Formally Undecidable Propositions of Principia Mathematica and Related Systems," see Ernest Nagel and James R. Newman, *Gödel's Proof* (New York: New York University Press, 1960). [Trans.]

thus introduced into the realm of logic and mathematics soon gave rise to an idea of "construction," just as in biology the hierarchy of properties suggested the idea of "evolution," for what could be more reasonable than that a relatively "weaker" structure have more elementary instruments and that to growing "power" there should correspond increasingly more complex "organs"?

This idea of "construction," which we must now try to explain, is incompatible with any simplistic theory of mind or intellectual activity.

Gödel showed that the construction of a demonstrably consistent relatively rich theory requires not simply an "analysis" of its "presuppositions," but the construction of the next "higher" theory! Previously, it was possible to view theories as layers of a pyramid, each resting on the one below, the theory at ground level being the most secure because constituted by the simplest means, and the whole firmly poised on a self-sufficient base. Now, however, "simplicity" becomes a sign of weakness and the "fastening" of any story in the edifice of human knowledge calls for the construction of the next higher story. To revert to our earlier image, the pyramid of knowledge no longer rests on foundations but hangs by its vertex, an ideal point never reached and, more curious, constantly rising! In short, rather than envisaging human knowledge as a pyramid or building of some sort, we should think of it as a spiral the radius of whose turns increases as the spiral rises.

This means, in effect, that the idea of *structure* as a system of transformations becomes continuous with that of *construction* as continual formation. Though this may appear puzzling at first sight, the reason for it is really quite simple. From Gödel's conclusions there follow certain important insights as to the limits of formalization in general; in particular, it has been possible to show that there are, in addition to formalized levels of knowledge, distinct "semi-formal" or "semi-intuitive" levels, which wait their turn, so to say, for formalization. The limits of formalization are not laid down once and for all, like the walls of China, but, instead, are "moveable" or "vicarious." J. Ladrière neatly sums up what is here involved in the following statement: "We cannot survey all the operations open to human thought at one glance."[12] Though this is a sufficiently exact first approximation, it should be supplemented by the reminder, first, that the number of operations open to human thought is not fixed and may, for all we know, grow; second, that this very capacity for surveillance becomes so greatly altered as the mind develops (as we know from psychogenetic studies) that it is not unreasonable to hope that its reach could be extended.

If, now, we appeal to our earlier notion of the correlativity of form and content (see Section 7), the limits of formalization can, more simply, be understood as due to the fact that there *is* no "form as such" or "content as such," that each element—from sensory-motor acts through operations to theories—is always simultaneously form to the content it subsumes and content for some higher form. Elementary arithmetic, for example, is no doubt from one perspective a "form," but from the perspective of transfinite arithmetic it is a "content," namely, the "denumerable." At each level, formalization of a

[12] *Dialectica,* XIV (1960), 321.

given content is limited by the nature of this content. Relative to concrete acts, "natural logic" is a "form," but one which can, in turn, be "formalized," though formalization cannot be carried very far. Again, intuitive mathematics is, from one point of view, a form, which can likewise become "formalized"; formalization here can be carried much further; nevertheless, as before, it requires "enrichment."

Human activity at every level seems to present us with "forms" (sensory-motor patterns, perceptual schemes among them, being one such variety of forms). Should we, then, end our account with the proclamation "Everything is structure" and let it go at that? No, for though it is true that everything can *become* structured, the difference in modality is all-important. Structure in the technical sense of a self-regulating system of transformations is not coincident with form: even a stack of pebbles can be said to have *form* (there are "bad" as well as "good" forms; see Section 11), but this mere heap cannot become a *structure* unless we place it in the context of a sophisticated theory by intercalating the system of all its "virtual" movements. We are thus brought to physics.

III

PHYSICAL AND BIOLOGICAL STRUCTURES

9. *Physical Structures and Causality*

It was imperative to begin with an examination of structuralist developments in mathematics and logic. But, since structuralism is chiefly known as the theoretical position responsible for avant-garde movements in the human sciences, we may well be asked why we should want to linger over physics. We answer, because we want to know, and cannot know *a priori,* whether man, nature, or both are the source of structure; and it is on the terrain of physics that they come together.

For a long time the physicist's ideal of science was to measure the phenomena, to establish quantitative laws, and to interpret these laws in terms of certain functionally interdependent conceptions like acceleration, mass, work, energy, and so on, so defined as to preserve the fundamental conservation principles which gave coherence to scientific theory. If, in considering this classical stage of physics, we are to speak of structures at all, we must reserve the word primarily for the major theories within whose frame relations were adjusted into relational systems: in Newton, inertia, equality of action and reaction, force as product of mass and acceleration; in Maxwell, the reciprocity of electrical and magnetic processes. In contemporary physics, the situation is quite different: the physics of the *Principia* (and therewith of "principles" as Newton understood them) has been shaken; research has been extended to "extreme" levels above and below phenomena; customary ways of thought have been overturned (consider the quite unexpected subordination of mechanics to electromagnetism). Measurement must now proceed within the context of a theory of measurement (the most sensitive point of contemporary physics), and structure, conceived as the set of *possible* states and transformations of which the system that actually obtains is a special case, moves to the forefront as prior to measure. The actual is now interpreted or explained as an instance of the possible.

The central problem which these developments in physics raise for structuralism is how causality is to be understood; more exactly, how the formal, that is, mathematical or logical, structures in terms of which physical laws are explained are related to the structures ascribed to the real. Of course, if we go along with the positivists and view mathematics as simply a language, the problem does not arise; science reduces to description. But with the admission of logical or mathematical structures of the transformational sort, the problem becomes unavoidable. Is it these formal transformations which ultimately account for change and persistence in the observed facts? Or are such transformational systems merely an interior reflection of the mechanisms of an objective physical causality that exists apart from us?[1] Or, finally, must we, to make sense of the fact that we are in possession of knowledge of nature, allow for some sort of permanent tie, though not of identity, between "external" structures and the structure of "our" operations? If there is such a connection, we should find it in evidence in "intermediate" regions: biological structures and our own sensory-motor acts should exhibit it in its efficacy.

Returning to physics, what speaks in favor of this last alternative is that logico-mathematical deduction of a set of laws is not sufficient for their explanation, at least not so long as deduction remains formal; explanation re-

[1] As examples of the first and second alternative we may take Meyerson's and Brunschvicg's doctrines, Meyerson conceiving of causality as *a priori*, because reducible to the "identifying of the diverse," Brunschvicg defining causality by the formula "there is a universe" (in the sense of relativity theory). But obviously Meyerson can account only for "conservations" and must relegate all "transformations" to the realm of the irrational, though it is precisely because we want to understand change that we call on the principle of causality. And Brunschvicg is saddled with the problem that, on his theory, there can be no distinction between "operations of thought" and "laws of nature," or, rather, the former are absorbed into the latter, so that arithmetic, for example, becomes a "mathematico-physical" discipline (this despite all that might be said about Brunschvicg's "idealism"!). If true, this hypothesis should be capable of psycho-biological verification.

quires, besides, that something be supposed to underlie phenomena and that these hypothetical objects really act upon one another. Now the striking fact is that frequently the action of such inferred entities resembles our own operations, and it is precisely to the extent that there is such a correspondence between inner and outer that we feel we "understand." Thus, understanding or explaining is not just a matter of applying our operations to the real and finding that "it can be done." Such "application" does not break through to causes; it keeps us within the realm of laws. Causal explanation requires that the operations that "fit" the real "belong" to it, that reality itself be constituted of operators.[2] Then and only then does it make sense to speak of "causal structures," for what this means is the *objective* system of operators in their effective interaction.

From this perspective, the steady agreement between physical reality and the mathematical theories employed in its description is of itself amazing, since the mathematics so often antedates its physical application, and, even when the mathematical apparatus is devised to fit certain newly found facts, it is nevertheless never derived from these facts but constructed as a deductively elaborated match for them. This harmony between mathematics and physical reality cannot, in positivist fashion, be written off as simply the correspondence of a language with the objects it designates. Languages are not in the habit of forecasting the events they describe; rather, it is

---

[2] We are here generalizing what is already being done in microphysics, where interdependent "operators" replace observed magnitudes.

a correspondence of human operations with those of object-operators, a harmony, then, between this particular operator—the human being as body and mind—and the innumerable operators in nature—physical objects at their several levels. Here we have remarkable proof of that pre-established harmony among windowless monads of which Leibniz dreamt, or—should the monads be "open"—the most beautiful example of biological adaptation that we know of (because physico-chemical and cognitive at the same time).

What holds for operations in general holds also for the most notable operational structures. Group structures, for example, have, as is well known, very wide application in physics—at the microphysical level, and all the way to relativistic celestial mechanics—a fact which has an important bearing on our question concerning the relations between the subject's operational structures and objective external operators. Three cases can here be made out. First, the group may have heuristic value, though the transformations it involves cannot be realized physically; the quaternary PCT (P standing for parity, the transform of a configuration into its symmetrical mirror image; C standing for charge, the transform of a particle into its anti-particle; and T standing for the inversion of the direction of time) can serve as an example. Second, the transformations, without being tantamount to physical processes that would exist apart from the physicist, are nevertheless results of either the material action of the experimenter handling his instruments or the coordination of the several pointer readings of differently placed observers; one of the Lorentz transformations provides an instance—to coordinate the points of view of two observers of different velocity, it introduces a change of reference point. In both these cases the group transformations are operations of the subject, but, as the case just cited shows, they can be real manipulations of the system under study.

This brings us to the last case, where the group transformations are physically realized apart from the manipulative activity of an experimenter or, though merely potential or "virtual," have a physical significance. Whenever the parallelogram of forces represents the compounding of forces that enter into combination of themselves, we meet with a case of this last and most interesting type. To see how the group concept enters here it should be recalled, first, that to obtain a condition of equilibrium it is sufficient to invert R, the resultant of two given forces (that is, for any F_1 and F_2 whose resultant is R, the resultant of F_1, F_2, and R'—a force equal and opposite to R— $= 0$); second, that equilibrium states are explained in terms of "compensation" for all "virtual work" compatible with the system's connections; together these two ideas amount to one vast explanatory "structure" based on the group concept.

Max Planck, whose part in the creation of quantum physics is well known, but who, as is also known, had reservations about some of the ideas given currency by his researches, maintained that side by side with efficient causality there is a principle of least action, and that physical phenomena obey the latter as strictly as the former. On his view, this principle derives from a "final cause which . . . turns the future or, more exactly, some determinate end, into that from which the processes leading to the end themselves evolve."[3] While granting "operator" status to inferred entities, we may not want to go as far as Planck, who endows the photons in the light rays that reach us from the stars with the capacity of behaving like "beings possessed of reason" on account of their following a minimum optical path despite all the refractions undergone in traversing the layers of the atmosphere; but we must inquire how Fermat's integral, the shortest path, is determined in such a case. Here again, as in accounting for states of equilibrium, it is by placing the real in the context of possible transformations that the solution is found, namely, that according to which the possible variant paths in the neighborhood of the real trajectory tend to cancel one another.

The role of the possible stands out, finally, in probabilistic explanations. Probability being defined as the ratio of favorable to possible cases, explaining the second principle of thermodynamics in terms of an "increase of probability" (entropy) amounts (despite the fact that there is here a type of irreversibility which runs counter to group structure) to determining a "structure" by forming the set of possibles from which the real is then to be selected.

In sum, there are physical structures which, though independent of us, correspond to our operational structures, especially in sharing the quasi-intellectual trait of covering the possible and locating the real within a system of virtuals. This kinship between causal and operational structures is quite what one would expect so long as

[3] Max Planck, *L'image du monde dans la physique moderne* (Gonthier, 1963), p. 130.

explanation involves models (which are at least in part artificial constructs) or relates to those special situations in microphysics where physical processes cannot be detached from the activity of the experimenter; but when the causal structure is quite external, its correspondence with operational structures poses a problem. The simplest explanation is to recall: (1) that it is in our own action that we first discovered causality, not the action of a self in the metaphysical sense of, say, Maine de Biran, but the intentional sensory-motor action whereby the young child first becomes aware of the transmission of movement and the role of push and resistance; (2) that action in this sense is the source of operations as well. Not that it "contains" them *ab initio,* it does not, just as it does not "contain" the whole of causality, but that its general coordinations involve certain elementary structures sufficient to serve as point of departure for reflective abstraction and, eventually, more complex constructions. We are thus brought to biological structures.

10. *Organic Structures*

The living organism is both a physico-chemical system among other such systems and the source of a subject's activities. If, then, as we have maintained throughout, a structure is a systematic whole of self-regulating transformations, the organism is, in a way, the paradigm structure: if we knew our own organism through and through, it would, on account of its double role of complex physical object and originator of behavior, give us the key to a general theory of structure. But after centuries of simplistic reductionism on the one hand and of a vitalism more verbal than explanatory on the other, biological structuralism is as yet only in its beginnings.

All problems of reduction are of interest to the structuralist, but those encountered in the effort to reduce vital phenomena to physico-chemical ones are of special interest, both because the problems are here most acute and because, as was just said, the organism is in a manner the prototype structure. The fundamental principle of reductionist programs has always been: Knowing phenomena $A, B, C \ldots$ in the inorganic world, we need only compound their sum or product to arrive at an understanding of organisms. Descartes' "animal machines" and the still widely held theory of evolution by fortuitous variation cum selection (selection being tagged on to make good on the facts for which fortuitous variation cannot account) are only two of the most erroneous in a long series of "mechanist" doctrines based on this principle. All such theories overlook two facts of capital importance: first, that progress in physics never takes the form of simply "adding on" new information—new discoveries M, N always lead to a complete recasting of preceding knowledge A, B, C while leaving room for some future discovery of Q, R, S; second, that even in physics attempts to reduce the complex to the simple, for example, electromagnetic to mechanical phenomena, lead to syntheses in which the more basic theory becomes enriched by the derived theory, and the resulting reciprocal assimilation reveals the existence of structures as distinct from additive complexes. We can, therefore, be quite relaxed about the prospect that living phenomena will one day become reduced to physico-chemical ones; here, as in physics, reduction will not mean impoverishment but such transformation of the two terms connected as benefits both.

Vitalism has always opposed anti-structuralist attempts to reduce and simplify by calling on ideas of wholeness, internal and external finality, and so forth, but so long as their mode of operation is left unspecified, these are not structuralist concepts. The same holds for the theory of emergence defended by Lloyd Morgan and others; to note the existence of wholes at different levels and to remark that at a given moment the higher "emerges" from the lower is to locate a problem, not to solve it. Moreover, though in their opposition to mechanist reductions of organisms to objects the vitalists rightly stressed that the organism is, if not the subject, at least the source of the subject, they never developed an adequate theory of the subject but were content to describe it either in common-sense introspective terms or, as in the case of Driesch, in terms derived from Aristotelian metaphysics.

In this connection it is interesting to note that the very first attempt to introduce an explicitly structuralist perspective into biology, namely, the "organicism" of L. von Bertalanffy,[4] was inspired by work in experimental psychology concerned with perceptual or motor schemes (*Gestalten*). In other words, structuralism made its entry into biology when the adequacy of available accounts of the activity of the subject became questioned. But though

---

[4] See Ludwig von Bertalanffy, *Problems of Life: An Evaluation of Modern Biological and Scientific Thought* (New York: Harper Torchbook ed., 1960), pp. 205ff., for a list of "publications pertinent to the foundation of the organismic conception." [Trans.]

the work of this theoretician in biology—an attempt to develop a "general theory of systems"—is of undeniably great interest, internal developments in comparative physiology, causal embryology, genetics, evolutionary theory, ethology, etc., tell us more about the contemporary structuralist orientation of biology.

Building on the work of Claude Bernard, physiology has long employed a concept that is of prime importance from the structuralist perspective, namely, the notion of "homeostasis," introduced into biology by Cannon. The idea of homeostasis, of the organism as so regulating internal conditions as to preserve a permanent state of equilibrium, prepares us to see the organism in its entirety as a self-regulating system. In three respects organic self-regulation goes beyond those known physical mechanisms of equilibration of which Le Chatelier's principle of "partial compensation" for any disturbance of equilibrium is the most notable.

First, we find that the regulation of an organism's structure, which is at first due to a general self-regulation, is later ensured by differentiated organs of regulation. Thus, the multiple factors entering into the coagulation of the blood stem, according to Markosjan, from some spontaneous regulation that is phylogenetically very early (probably as early as the coelenterates) and which, with the development of the hormonal system, became subject, first to one organ of regulation, and later, with the development of the nervous system, to yet another.

Second, and because of the facts just cited, a living structure's functioning is always tied to the functioning of the total organism, the subordinate function (in the biological sense of the word, of course) being definable in terms of the substructure's relation to the organism's total structure. In biology this tie between function and structure can hardly be denied; but in the realm of cognitive psychology there are some who think of structuralism as excluding all functionalism, a view we shall have to take up later.

Third and, be it noted, in direct connection with this functional character of organic structures, the latter present an aspect which physical structures lack (except in so far as the inquiring physicist must sometimes be considered part of such structures): they take account of *meanings*.[5] The behavior of the living subject depends upon quite explicit meanings: instinctual structures, for example, function in terms of all sorts of hereditary "clues"—the IRM's, "innate releasing mechanisms," of the ethologists. But meanings are implicit in all functioning, even the specifically biological distinction between normal and abnormal conditions depends on them; for example, when, at birth, there is danger of suffocation, the coagulation of the blood immediately gives rise to regulation through the nervous system.

The concept of homeostasis is, however, not limited to physiology. One of the most important victories of contemporary biological structuralism is the rejection of gene complexes as previously conceived: while these were earlier viewed as aggregates of isolated genes, they are today understood as gene systems, the individual genes no longer performing, as Dobzhansky put it, "as soloists, but as members of an orchestra." In particular, there are

[5] See Piaget's *The Origins of Intelligence in Children* (New York: Norton, 1963), pp. 189ff., where Piaget briefly explains his concept of meaning. [Trans.]

"regulator genes" which cause the concerted action of several genes in the production of a single trait or a single gene's affecting several traits, and so on. And the genetic whole is no longer the gene complex of the individual but the "population," which is not a mere "mix" but a combination of races furnishing a genetic "pool" that is in "genetic homeostasis," i.e., so equilibriated as to increase the chances of survival. Dobzhansky and Spassky have verified this hypothesis of genetic homeostasis by mixing several known breeds in a "population cage" and studying their offspring after several generations. It should also be mentioned that the fundamental mechanism of variation is, on present-day theory, no longer mutation but genetic "recombination," new hereditary structures being chiefly formed by a reassortment of genes.

In embryology the structuralist tendencies that were first given currency by the discovery of "organizers," structural regulations, and regenerations, have now become accentuated through the work of C. H. Waddington,[6] which introduces a notion of "homeorhesis" (see Waddington, p. 32) according to which embryological development involves a kinetic equilibration whereby deviations from certain necessary paths of development ("créodes") are compensated for. More important still, Waddington has shown that environment and gene complex interact in the formation of the phenotype, that the phenotype is the gene complex's response to the environment's incitations, and that "selection" operates, not on

[6] C. H. Waddington, *The Strategy of the Genes: A Discussion of Some Aspects of Theoretical Biology* (New York: Macmillan, 1957). [Trans.]

the gene complex as such, but on these responses. By insisting on this point, Waddington has been able to develop a theory of "genetic assimilation," i.e., of the fixation of acquired characteristics. Roughly, Waddington views the relations between the organism and its environment as a cybernetic loop such that the organism selects its environment while being conditioned by it. What this means is that the notion of structure as a self-regulating system should be carried beyond the individual organism, beyond even the population, to encompass the complex of milieu, phenotype, and genetic pool. Obviously, this interpretation of self-regulation is of the first importance for evolutionary theory.

Just as there still are embryologists who remain wedded to an entirely preformational view of ontogenesis and who, accordingly, deny all epigenesis (restored to its plain sense by Waddington), so it has occasionally been maintained of late that the entire evolutionary process is predetermined by the combination established by the constituents of the DNA molecules. Thought through to the end, preformational structuralism simply cancels the idea of evolution. Waddington, by reestablishing the role of the environment as setting "problems" to which genotypical variations are a response, gives evolution the dialectical character without which it would be the mere setting out of an eternally predestined plan whose gaps and imperfections are utterly inexplicable.

These advances in contemporary biology are all the more valuable to structuralism because, joined to ethology (the comparative study of animal behavior), they furnish the basis for psychogenetic structuralism. Ethologists have shown that there is a complex structure of instincts. We may even go so far as to speak of a "logic of instincts" whose several "levels" can be subjected to analysis, the hierarchy of instincts thereby becoming a "logic" of organs or organic instruments antedating the logic of acts (organic activities that are not "genetically programmed") or of manufactured instruments. But what is no less essential is that contemporary ethology tends to show that all learning and remembering depend upon antecedent structures (conceivably the DNA and RNA themselves). Thus, the contacts with experience and the fortuitous modifications due to the environment on which empiricism modeled all learning do not become stabilized until and unless assimilated to structures; these structures need not be innate, nor are they necessarily immutable, but they must be more settled and coherent than the mere gropings with which empirical knowledge begins.

In short, biological wholes and self-regulating systems, though "material" and of physico-chemical content, enable us to understand the connection between "structures" and "the subject," because it is the organism which is the latter's source. If man, as Michel Foucault puts it,[7] is only "a kind of rupture in the order of things," to which has corresponded (but for less than two centuries) "a mere wrinkle in our knowledge," it is nevertheless worth remembering that this rupture and this wrinkle are the product of a great upheaval—but a well-organized one— which is constituted by life as a whole.

[7] Michel Foucault, *Les mots et les choses* (Paris: Gallimard, 1966), p. 15.

IV

PSYCHOLOGICAL STRUCTURES

11. *Gestalt Psychology and the Beginnings of Structuralism in Psychology*

The idea of structure may be said to have been introduced into psychology toward the beginning of this century when the Würzberg school of "cognitive psychology"[1] (like Binet in France and Claparède in Switzerland) took a stand against associationism. It is remarkable that, at just about the same time, K. Bühler, by strictly experimental means, demonstrated the subjective aspects of the structure which has since then figured so prominently in phenomenology—"intention" and "significance" (the phenomenological counterparts to our "transformation" and "self-regulation" as employed in the "objective"

---

[1] See, for example, Oswald Külpe's later work. [Trans.]

definition of structure offered in Section 1). Bühler showed not only that judgment is an act of unification, a thesis immediately endorsed by all anti-associationists, but also that there are ascending degrees of complexity in thought. He differentiated three such stages or levels of complexity: the first he called *Bewusstheit* ("consciousness," that is, thought as independent of images and as ascribing significance), the second *Regelbewusstheit* (consciousness of the rules involved in relational structures, and so on), and the third *intentio,* by which he meant the deliberate synthetic act "intent upon" the construction of a whole, that is, a system of thought "at work."

However, instead of attempting to uncover the psychogenetic and biological roots of thought, the Würzberg school's studies were limited to the already formed adult intelligence (moreover, as need hardly be pointed out, the "adult" studied by psychologists is always picked from among his assistants or students); no wonder, then, that the only structures brought to light were logical structures, and that the inescapable conclusion was that "thought is the mirror of logic"; if they had turned their attention to the *genesis* of intelligence, these terms would surely have been reversed.

But the most spectacular form of psychological structuralism was undoubtedly the theory of Gestalten which grew out of the convergent researches of Wolfgang Köhler and Max Wertheimer and which became extended to social psychology by Kurt Lewin[2] and his pupils.

Though Gestalt psychology developed originally in the

[2] See Chapter VI.

ambience of phenomenology, it retained only the phenom-
enologists' emphasis upon interaction between subject
and object[3] and was resolutely naturalistic in its orientation. Köhler had, after all, been trained as a physicist, and
it was physics which suggested the notion, fundamental
to him and other Gestalt psychologists, of the "field." As
we shall see, stimulating as the introduction of field
models was initially, the dominant role assigned to them
had some ill effects.

A field of force, such as an electromagnetic field, is an
organized whole in that the net force exerted upon anything in that field depends upon the direction and inten-
sity of all the forces; but the compounding of forces is a
practically instantaneous affair, so that, though we may
bring it under the rubric of transformation, the transformations here involved are quasi-immediate. Now even
if we restrict attention to the nervous system and poly-
synaptic fields, electric impulses are not by any means
transmitted instantaneously (they travel at a rate of 3 to
9 cycles/second for waves δ to α). And granting that,
upon reception of the nerve impulses, the organization of
perception is extremely rapid, it would not be legitimate
to infer that this holds for all Gestalten. It was the pre-
occupation with field effects which led Köhler to his
excessively narrow interpretation of intelligence—only
an act of "immediate insight" being viewed as "intelligent," as if the exploratory groping antecedent to the
final intuition were not itself intelligent. Most important,
it is undoubtedly the field concept that must be blamed

[3] Interaction of subject and object is, of course, emphasized
just as much by Brunschvicg, by dialectical modes of thought in
general.

for the Gestalt theorists' slighting of all functional and psychogenetic considerations and, ultimately, of the agency of the subject.

Nevertheless, precisely because conceived in this way, the psychological Gestalt represents a type of structure that appeals to those who, whether they acknowledge it or not, are really looking for structures that may be thought "pure," unpolluted by history or genesis, functionless and detached from the subject. Philosophy lends itself to the construction of such essences; here invention is not hemmed in. But such purity is hardly to be met with in the domain of empirical reality, unless the Gestalt hypothesis should turn out to be true. We must, then, carefully examine the arguments in its favor.

The central idea of Gestaltist structuralism is the idea of wholeness. As early as 1890, Christian von Ehrenfels showed that complex perceptual units like a melody or a person's physiognomy have perceptual qualities which accrue to them as configurations (Gestalten); they have *Gestaltqualitäten*.[4] This is why a melody, transposed into another key, all the individual notes changed, will nevertheless be recognized as "the same melody." Ehrenfels viewed such *Gestaltqualitäten* as perceptual realities supervening and imposed on sensation. The originality of the theory of Gestalten consisted in contesting the existence of "sensation as such": Whereas on the associationist theory sensations are the "given," structured perceptual wholes arising somehow from them, the Gestalt psychologists maintained that what is given is

---

[4] See Wolfgang Köhler, *Gestalt Psychology* (New York: Mentor, 1947), pp. 102ff. [Trans.]

always from the start a whole, a structure within which
sensations figure only as elements. The perceptual whole
is the datum that calls for explanation; this is where the
field hypothesis comes in. On this hypothesis, the incom-
ing nerve impulses do not strike the brain one by one,
sequentially, for by the mediation of the electric field of
the nervous system they almost instantaneously give rise
to "forms" of organization. But the laws governing such
organization remain to be discovered.

Since the elements in a field are always subordinated
to the whole, every local modification engendering a re-
fashioning of the ensemble, the first law of perceptual
totalities is that the whole, over and beyond its having
qualitative features of its own, has a quantitative value
different from that of the sum of its parts. That is to say,
the law of composition for perceptual wholes is non-
additive. As Köhler tells us explicitly in *Die physischen
Gestalten in Ruhe und im stationären Zustand*, the compounding of mechanical forces lacks Gestalt character
precisely because such compounding *is* additive. This
first law is easily verified: a divided space appears larger
than an undivided one; a leaden bar hefted by itself will
feel heavier than that same bar when mounted on an
empty box (box and bar together forming a simple shape
and being painted the same color), and so on.

The second law is that perceptual totalities tend to
take on the "best" form possible, "good" forms being
simple, regular, symmetrical, closely packed, and so
forth. On the field hypothesis, the prevalence of good
forms is a consequence of physical principles of equilibration and least action, the same principles that

account for the sphericity of soap bubbles (maximum volume for minimum surface).

There are many other important and amply verified Gestalt theoretical laws—the figure/ground law, the law that boundaries are always perceived as belonging to the figure, not the background, etc. But the two discussed will suffice for our purposes.

Before we go on, we should stress the importance of this notion of equilibration, which enables us to dispense with an archetypal explanation for the prevalence of good forms. Since equilibration laws are coercive, they suffice to account for the generality of such processes of form selection; heredity need not be called in at all. Moreover, it is equilibration which makes Gestalten reenter the domain of structure as circumscribed by us in Section 1, for whether physical or physiological, equilibration involves the idea of transformation within a system and the idea of self-regulation. Gestalt psychology is therefore a structuralist theory more on account of its use of equilibration principles than because of the laws of wholeness it proposes.

On the other hand, the adequacy of the field hypothesis, with its various anti-functionalist consequences, may well be doubted. Thus, Piéron has shown that if two visual stimuli for the perception of apparent motion are presented separately to one eye, no such motion is registered, because the immediacy of the circuit between the two cerebral hemispheres assumed by Gestalt theory does not in fact exist. The fact that it is possible to train perception in a variety of ways is hardly compatible with a physicalist interpretation in terms of fields. E. Brunswick has demonstrated that what he calls "empirical Gestalten" must be differentiated from "'geometrical Gestalten'": when, with the help of a tachistoscope, a shape midway between a hand and a rather symmetrical five-pointed figure is briefly exposed to view, only half of the adults "correct" the model in the direction of "good geometric form"; the other half interpret it as a hand (the relevant "empirical Gestalt"). Now if, as Brunswick contends, perception is modified by experience—depends, that is, upon the relative frequency with which the empirical prototype has been met—the structuring of perception is governed by functional and not only by physical laws. Indeed, Wallach, Köhler's chief collaborator, felt obliged to admit that memory plays a role in perceptual structuring.

Our own researches,[5] conducted with the help of numerous collaborators, have shown that perception develops with maturation and that, granting the existence of field effects (but understood in the sense of a field of visual attention), "perceptual acts" like relating by quasi-deliberate exploration, comparing, and so on (which modify the perceived Gestalten noticeably) must also be taken into account. For example, when we study visual "exploration" of figures by recording eye movements, we find that eye movements become increasingly better coordinated and adjusted with maturation. As for field effects, their quasi-immediate interaction appears to be due to probabilistic mechanisms of "encounter" between

[5] Jean Piaget, *The Mechanisms of Perception*, trans. G. N. Seagrin (New York: Basic Books, Inc., 1969).

the parts of the receptive organ and portions of the perceived figure; "couplings" or correspondence between these encounters are particularly important; and from this probabilistic schema a law coordinating the various known optic-geometrical illusions can be derived.

In short, even perception calls for a subject who is more than just the theatre on whose stage various plays independent of him and regulated in advance by physical laws of automatic equilibration are performed; the subject performs in, sometimes even composes, these plays; as they unfold, he adjusts them by acting as an equilibrating agent compensating for external disturbances; he is constantly involved in self-regulating processes.

What holds for perception holds all the more for motor activity and intelligence, which the Gestaltists tried to subordinate to laws of Gestalt—particularly perceptual Gestalt—formation. In his otherwise admirable book, *The Mentality of Apes,* Köhler presents understanding as the sudden reorganization of the perceptual field in terms of better forms; and Wertheimer tried to reduce even syllogizing or mathematical reasoning to processes of restructuring governed by Gestalt laws. But these interpretations, which proceed by extending the field hypothesis, meet with two obstacles. First, though logico-mathematical structures are without the least shadow of doubt subject to laws of wholeness (see Sections 5 through 7), they are not Gestalten, since their laws of composition are strictly additive (2 plus 2 make exactly 4, even though, or precisely because, this addition "participates in" the group laws). Second, the sensori-motor or intelligent subject is active and himself constructs his structures, by operations of reflective abstraction which only rarely and in exceptional cases have any noticeable resemblance to perceptual figuration. But this raises a central problem of structuralism which we must now examine more closely.

12. *Structure and the Genesis of Intelligence*

All sorts of origins can be ascribed to structures. They may be viewed as given as such, in the manner of eternal essences; or as surging up inexplicably in the course of that capricious history of which Michel Foucault offers us the archaeology; or as derived from the physical world in the manner of Gestalten; or, finally, as somehow dependent upon the subject. Nevertheless, there are only a finite number of such alternatives, and these may be said to cluster into three main groups: the hypotheses in the first group tend toward an innatism reminiscent of predetermination, with this difference, that the hereditary origins are viewed as biological, so that the problem of their first formation is unavoidable; those in the second group tends toward a theory of contingent "emergence" (of which Foucault's "archaeology" is a special version); and finally there are the constructivist accounts. So there are just three solutions: preformation, contingent creation, or construction. "Derivation" from experience does not furnish an additional solution since, in spelling out what this might mean, one would have to assume either that experience obtains its structure from antecedent conditions, or that it gives direct access to external structures which must, accordingly, themselves be preformed in the external world.

Since the idea of contingent emergence is pretty nearly incompatible with the idea of structure (we shall return to this in Section 21), in any case, with logico-mathematical structures, the real problem is to decide between predetermination and construction. And inasmuch as structures are closed and autonomous wholes, preformation seems, at first sight, the only plausible account of their origin; this is why Platonist tendencies are constantly reborn in mathematics and logic and why a certain static structuralism is so successful with authors who are enamored of absolute beginnings and theories unpolluted by history or psychology. On the other hand, as transformational systems which derive from one another by more or less abstract "genealogies," and as "operational" in their paradigm instances, structures suggest the formational hypothesis, and self-regulation seems to call for self-construction.

All studies of the formation of intelligence run into this dilemma; the facts themselves force it upon us when, for example, we try to explain how the child eventually comes to master logico-mathematical structures. Either he discovers them ready-made—but obviously he does not learn of their existence as he learns about colors or falling bodies, and to be taught about them, whether at school or at home, he must already be in possession of certain minimal instruments of assimilation, which themselves partake of such structures (that this holds for language teaching as well will be shown in Section 17). Or else he "constructs" them—but he is by no means free to draw them up at his pleasure (like the rules of a game or the figures in a picture), so the question of how and why this construction yields *necessary* results remains. Why does it look "as if" the results were "predetermined"?

Now observation and experiment show as clearly as can be that logical structures *are* constructed, and that it takes a good dozen years before they are fully elaborated; further, that this construction is governed by special laws, laws which do not apply to any and every sort of learning. Through the interplay of reflective abstraction (see Section 5), which furnishes increasingly complex "materials" for construction, and of equilibration (self-regulation) mechanisms, which make for internal reversibility, structures—*in being constructed*—give rise to that necessity which a priorist theories have always thought it necessary to posit at the outset. Necessity, instead of being the prior *condition* for learning, is its *outcome*.

Of course, human structures do not arise out of nothing. If it be true that all structures are generated, it is just as true that generation is always a passing from a simpler to a more complex structure, this process, according to the present state of our knowledge, being endless. So there are certain givens from which the construction of logical structures takes off, but these "data" are not primordial in any absolute sense, being merely the starting point for our analysis, nor do they "contain" what is, in the course of construction, "derived" from and "based" on them. We called these initial structures behind which we cannot go "general coordinations of actions," meaning to refer to the connections that are common to all sensori-motor coordinations. Elsewhere[6] we have analyzed the various stages of sensori-motor development—from the organism's first spontaneous movements and reflexes (the latter undoubtedly stabilized selections from among the former), or from reflex-complexes like the sucking movements of the neonate, through acquired habits, and up to the beginnings of sensori-motor or practical intelligence. Here a global account will have to suffice. Now all such behavior that has innate roots but becomes differentiated through functioning contains, we find, the same functional factors and structural elements. The functional factors are *assimilation,* the process whereby an action is actively reproduced and comes to incorporate new objects into itself (for example, thumb sucking as a case of sucking), and *accommodation,* the process whereby the schemes of assimilation themselves become modified in being applied to a diversity of objects. The structural elements are, essentially, certain *order* relations (the order of movements in a reflex act, in a habitual act, in the suiting of means to end), *subordination schemes* (the subordination of a relatively more simple schema like grasping to a relatively more complex one like pulling) and *correspondences* (such as are involved in what we have elsewhere called "recognitory assimilation"). As the primary assimilation schemes become mutually coordinated ("reciprocally assimilated"), certain equilibriated structures, those that make for a modicum of "reversibility," become established. Most striking among these are, first, the "practical" group

[6] *The Origins of Intelligence in Children* (New York: Norton, 1963). [Trans.]

of displacements, which corresponds to the displacement group described in Section 5 with its associated invariance condition, that is, the permanence of objects which pass out of the perceptual field and can be found again by reconstituting their displacements; second, the spatialized and objectivized form of causality which enters into intentional acts like getting at things by tugging at their supports or by using a stick, and so on. Even at this stage the child's behavior may be called intelligent, but his intelligence is entirely sensori-motor, does not involve representation, and is essentially tied to action and coordinations of action.

As soon as the semiotic function (speech, symbolic play, images, and such) comes on the scene and with it the ability to evoke what is not actually perceived, that is, as soon as the child begins to represent and think, he uses reflective abstractions: certain connections are "drawn out" of the sensori-motor schemata and "projected upon" the new plane of thought; these are then elaborated by giving rise to distinct lines of behavior and conceptual structures. The order relations, for example, which on the sensori-motor plane were altogether immersed in the sensori-motor schema, now become dissociated and give rise to a specific activity of "ranking" or "ordering." Similarly, the subordination schemes which were originally only implicit now become separated out and lead to a distinct classificatory activity; and the setting up of correspondence soon becomes quite systematic: one/many; one/one; copy to original, and so on.

In observing this kind of behavior we undeniably meet with the advent of logic, but we should note that this logic is limited in two essential respects: such ordering or classifying or setting up of correspondences does not involve reversibility, so that we cannot as yet speak of "operations" (since we reserved that term for procedures which have an inverse), and because of this, there are as yet no principles of quantitative conservation (a divided whole is not "equal to" the previous undivided whole, and so on). So we should view this stage of intellectual development as a "semi-logical" stage, in the quite literal sense of lacking one-half, namely, the inverse operations. Nevertheless, two fundamental notions, that of *function* and that of *identity*, are in evidence even at this stage. For example, if a child is shown a piece of string bent at a right angle and one "leg" *A* is progressively shortened, the child understands perfectly well that the other, *B*, thereby gradually becomes lengthened; only, for him this does not mean that the piece of string as a whole, *A* plus *B*, remains constant in size, because he estimates lengths ordinally, in terms of the order of terminal points; for him "longer" is the same as "farther away"; he does not count unit intervals. And though this piece of string does not, for him, have a constant length, it is nevertheless the "same" piece of string throughout. However rudimentary his understanding of "function" and "identity," they constitute structures in the sense of "categories" discussed in Section 6.

Between the ages of roughly seven and ten the child enters upon a third stage of intellectual development, which involves the use of operations, albeit only "concrete" operations, such as remain applied to things. He now arranges things in series and understands that in lining them up, say, in order of increasing size he is at the same time arranging them in order of decreasing size; the transitivity of relations like bigger than, and so on, which previously went unrecognized or was noted as a mere matter of fact, is now something of which he is explicitly aware. Classification is now accompanied by quantification of what is included. Multiplicative matrices are employed. Numbers are "constructed" by synthesizing seriation and inclusion; and measurement is "constructed" by synthesizing "partition" and order relations. Magnitude, previously understood only in the ordinal sense, becomes cardinal, and the conservation principles which earlier were lacking are now established. These various operations have the structure of "semigroups," incomplete because non-associative, or we may look upon them as semi-networks (with lower but without upper limits, or the reverse). Their chief limitation, from the perspective of adult intelligence, is that compounding proceeds by approximation and is not combinatorial.

Upon analyzing these structures, it is not hard to see that the more complex proceed from the simpler as reflective abstraction, which provides all the structural elements, and equilibration, which makes for operational reversibility, jointly come into play. So we are here in the presence of, and can even follow step by step, an activity of construction that yields genuine structures, structures which are already "logical" and which are nonetheless "new" when compared with those that preceded: the transformations definitive of the new structure grow out of the formative transformations, from which they differ only in equilibration.

But this is not the end of the story. A new set of reflective abstractions leads to the construction of new operations *upon* the preceding, though nothing new is added, except, again, a reorganization, this time, however, of major consequence. First of all, in generalizing his classifications, the subject arrives at that classification of classifications (an operation raised to the second power) which is what we mean by combinatory activity; this is what yields classes of subclasses and Boolean networks. Second, the coordination of inversions proper to the reversibility of semigroups and the reciprocities that belong to semigroups of relations engender the quaternary group INRC of which we spoke in Section 7.

Returning to the problem with which we started, we conclude that there is room for an alternative that falls between absolute preformation of logical structures on the one hand and their free or contingent invention on the other. Construction, in being constantly regulated by equilibration requirements (a self-regulation whose conditions become the more stringent as it steers toward an equilibrium that is mobile and stable at the same time), finally yields a necessity that is a non-temporal, because reversible, law. True, there will always be some to urge that all the "subject" does is to bring "virtual" structures which subsist from eternity together, and, since mathematics and logic are the sciences of the possible, it would not be inconsistent for the logician or mathematician to remain content with this sort of Platonism. But as soon as he breaks beyond the confines of his expertise and tries to develop an epistemology, he will have to ask himself where, exactly, this region of the virtual is to be located. To call on essences to furnish the virtual with its underpinnings is to beg the question. Nor can the physical world provide its habitation. It makes far better sense to assign the virtual a place in organic life, though obviously only on condition that it be clearly understood that general algebra is not "contained" in the behavior of bacteria or viruses. So what remains is, again, the constructivist hypothesis, and is it not quite plausible to think of the nature that underlies physical reality as constantly in process of construction rather than as a heap of finished structures?

13. *Structure and Function*

There are thinkers who dislike "the subject," and if this subject is characterized in terms of its "lived experience" we admit to being among them. Unfortunately, there are many more for whom psychologists are by definition concerned with "subjects" in just this individual "lived" sense. We do not ourselves know any such psychologists; if psychoanalysts have the patience to attend to individual cases in which the same conflicts and complexes show up again and again, it is once more with a view to discovering common mechanisms.

In any case, the "lived" can only have a very minor role in the construction of cognitive structures, for these do not belong to the subject's *consciousness* but to his operational *behavior*, which is something quite different. Not until he becomes old enough to reflect on his own habits and patterns of thought and action does the subject become aware of structures as such.

If, then, to account for the constructions we have described we must appeal to the subject's acts, the subject here meant can only be the epistemic subject, that is, the mechanisms common to all subjects at a certain level, those of the "average" subject. So average, in fact, that one of the most instructive methods for analyzing its actions is to construct, by means of machines or equations, models of "artificial intelligence" for which a cybernetic theory can then furnish the necessary and sufficient conditions; what is being modeled in this way is not its structure in the abstract (algebra would suffice for this), but its effective realization and operation.

It is from this point of view that structures are inseparable from performance, from functions in the biological sense of the word. The reader may have felt that, in including self-regulation or self-governing in our definition of structure, we bypassed the set of necessary conditions. Everyone grants that structures have laws of composition, which amounts to saying that they are regulated. But by what or by whom? If the theoretician who has framed the structure is the one who governs it, it exists only on the level of a formal exercise. To be real, a structure must, in the literal sense, be governed from within (in Section 12 we offered some examples of such self-government or self-regulation). So we come back to the necessity of some sort of functional activity; and, if the facts oblige us to attribute cognitive structures to a subject, it is for our purposes sufficient to define this subject as the center of functional activity.

But why postulate such a center? If structures exist and each is regulated from within, what role is left for the subject? Calling it the center of functional activity seems to amount to just that demotion for which we chided the Gestalt psychologists—the subject becomes the mere stage on which the various autonomous structures act out their predetermined role. Why not eliminate it altogether, as certain contemporary structuralists dream of doing?

If cognitive structures were static, the subject would indeed be a superfluous entity. But if it should turn out that structures tend to become connected in some way other than by a preestablished harmony among windowless monads, then the subject regains the role of mediator. It will either be the "structure of structures," the transcendental ego of a priorist theories, or, perhaps, more modestly, the "self" of psychological theories of synthesis (such as are proposed in P. Janet's first work, *L'automatisme psychologique;* Janet himself, on account of his interest in the dynamics of cognition, has extended this earlier theory in the functional and psychogenetic direction). Or, lacking this overarching power of synthesis, and having no structures at its disposal until it constructs them, the subject will, more modestly, but also more realistically, have to be defined in the terms we earlier proposed, as the center of activity.

At this point we should remind ourselves that the mathematicians have, in effect, settled the question as to the nature of the subject for us. Without their being aware of it, their results converge with those of psychogenetic analysis, which is quite amazing. There can be no "structure of structures" in the sense of a class of all classes, not only because of the well-known set-theoretical paradoxes, but also, and more profoundly, because of the limits of formalization (limitations which, in Section 8, we analyzed as resulting from the correlativity of form and content but which, as we now see, also depend upon the conditions of reflective abstraction; in the final analysis, this amounts to the same thing). In other words, the process of formalization itself is constructive; in the abstract it engenders a genealogy of structures while, concretely, the equilibration of these structures establishes psychogenetic filiations—between functions and semigroups, semigroups and groups, and so on.

In the construction proposed in Section 12, the function (in the biologist's sense of the word) chiefly credited for the formation of structures was "assimilation," our structuralist substitute for atomistic "association." Biologically considered, assimilation is the process whereby the organism in each of its interactions with the bodies or energies of its environment fits these in some manner to the requirements of its own physico-chemical structures while at the same time accommodating itself to them. Psychologically (behaviorally) considered, assimilation is the process whereby a function, once exercised, presses toward repetition, and in "reproducing" its own activity produces a schema into which the objects propitious to its exercise, whether familiar ("recognitory assimilation") or new ("generalizing assimilation"), become incorporated. So assimilation, the process or activity common to all forms of life, is the source of that continual relating, setting up of correspondences, establishing of functional connections, and so on, which characterizes the early stages of intelligence. And it is assimilation, again, which finally gives rise to those general schemata we called structures. But assimilation *itself* is *not* a structure. Assimilation is the functional aspect of structure-formation, intervening in each particular case of constructive activity, but sooner or later leading to the mutual assimilation of structures to one another, and so establishing ever more intimate inter-structural connections.

Before bringing Sections 12 and 13 to a close, we should mention the fact that, especially in America, our kind of structuralism is not unanimously endorsed. J. Bruner, for example, does not believe in "structures" or in "operations"; in his view, these are constructs ridden with "logicism" which do not render the psychological facts in and of themselves. He does credit the subject with cognitive acts and "strategies" (in the sense of von Neumann's theory of games).[7] But why, then, assume that such acts cannot become internalized and thereby turned into "operations"? And why must the subject's strategies remain isolated instead of becoming integrated into systems? Again, Bruner tries to account for the intellectual growth of the child in terms of the way in which he meets conflicts among the various modes of representation at his disposal—speech, images, action-schemes themselves. But if each of these models furnishes him with an incomplete and sometimes even distorting perception of reality, how can he resolve such conflicts unless he appeals either to some "copy" of reality or to "structures" as described by us, as coordinations of all instruments of representation? (We need hardly point out that, for a copy to serve as arbiter in case of conflict, we would have to trust it as a true copy, which would require some other means of access to its original than through it.) But on this second hypothesis, does not one of Bruner's representational schemes, namely language, become promoted

[7] See, for example, *A Study of Thinking* (New York: John Wiley, 1956; with J. J. Goodnow and G. A. Austin). [Trans.]

to privileged status, inasmuch as it has precisely that role of coordinating and structuring other modes of representation which we ascribed to structures? Must we not turn to linguistic structuralism to give the problems we discussed in this chapter a manageable form?

V

LINGUISTIC STRUCTURALISM

14. *Synchronic Structuralism*

Language is a group institution. Its rules are imposed on individuals. One generation coercively transmits it to the next, and this has been true for as long as there have been men. Any given form of it, any particular spoken language, derives from some earlier form, which in turn flows from some still more primitive form, and so on, indefinitely, without a break, all the way back to the one or more ancestral languages.

Every word in a language designates a concept, which constitutes its signification. The most resolute anti-mentalists, Bloomfield, for example, go so far as to maintain that the idea of concepts is completely reducible to that of a word's signification. More exactly, Bloomfield says that there are no concepts, that what is mistakenly so called is simply the signification of words. But surely this is one way of defining and granting existence to concepts.

The syntax and semantics of a language yield a set of rules to which any individual speaking that language must submit, not only when he wants to express his thought to others, but even when he expresses it "internally."

Language, in short, is independent of the decisions of individuals; it is the bearer of multi-millennial traditions; and it is every man's indispensable instrument of thought. As such, it appears to be a privileged domain of human reality, so it is only natural that it should sometimes be regarded as the source of structures which, on account of their age, generality, and power, are of special significance (that language and its structures far antedate science goes without saying).

Before taking up linguistic structures as the linguists understand them, we should note that there is an entire school of epistemology—logical positivism—according to which logic and mathematics furnish a "general syntax" and "general semantics"; seen from this perspective, the formal structures described in Chapter II are already "linguistic." This is not how we viewed them. We dealt with logical and mathematical structures as the products of "construction" and what we called "reflective abstraction," intellectual activities which ultimately derive, we said, from the processes of assimilation and accommodation whereby sensori-motor acts become coordinated with one another, and it is this activity of coordinating our own acts which we regarded as primary: applying to everything, it applies to acts of communication and exchange as well, consequently, to language. Linguistic structures lose none of their interest by being looked at in this way, but their relations to the structures of the signified become different. In whatever way the problem as to the relation between linguistic and logical structure be solved eventually, it is a fundamental problem for any general theory of structure.

Linguistic structuralism in the narrower sense goes back to Saussure, who showed that diachronic development is not the only process to be taken notice of in the study of a language, and that in fact the history of a word may give a seriously inadequate account of its meaning. In addition to its historical aspect language has a "systematic" aspect (Saussure did not use the term "structure"); it embodies laws of equilibrium which operate on its elements and which, at any given point in its history, yield a synchronic system. Since the basic relation in language is that between the sign and its meaning, and since meanings are relative to one another, the system is one of oppositions and differences; while it is synchronic because the meaning-relations are interdependent.[1]

Developed in opposition to the diachronic perspective

---

[1] See p. 140, *Course in General Linguistics:* "What diachronic linguistics studies is not relations between coexisting terms of a language-state but relations between successive terms that are substituted for each other in time," and pp. 99ff.: "Synchronic linguistics will be concerned with the logical and psychological relations that bind together coexisting terms and form a system in the collective minds of speakers." "Diachronic linguistics, on the contrary, will study relations that bind together successive terms not perceived by the collective mind but substituted for each other without forming a system." [Trans.]

of nineteenth-century comparative grammar, Saussure's structuralism, unlike the "transformational" structuralism of Chomsky and Harris today, was in essence synchronic. Since this has led a good many writers, not all of them linguists, to think of structures as inherently independent of history, we should dwell somewhat upon the reasons for the predominantly synchronic emphasis in early linguistic structuralism.

The first of these reasons is of a very general order and relates to the relative independence of laws of equilibruim from laws of development. Saussure, in elaborating on this point, drew his inspiration partly from economics, which in his day chiefly stressed the former (Walras' and Pareto's "general theory of equilibrium"); and it is of course true that economic crises may lead to radical shifts of value quite independent of antecedent price history—the price of tobacco in 1968 depends, not on its price in 1939 or 1914, but on the interaction of current market conditions. But Saussure might have drawn his argument for the relative autonomy of synchronic laws just as well from biology: an organ may change its function, and one and the same function may be exercised by different organs.

The second reason, which was perhaps psychologically the primary one, was the desire to give oneself over to the study of the immanent character of language without being distracted by historical considerations.

But it is the third reason which, for our present purpose, is the most important, for it relates to a circumstance peculiar to language, the arbitrariness of the verbal sign, which, since it is merely conventional, has no intrinsic nor, consequently, fixed relation to its meaning; there is nothing in the phonic character of the signifier to call forth the value[2] or content of the signified, and Saussure emphasized this point with systematic vigor. Jesperson already tempered this insistence upon the complete arbitrariness of the verbal sign, and more recently Jakobson threw further doubt upon it. Yet it should be noted that Saussure himself met these objections in advance by distinguishing between what he called the "relatively" and the "radically arbitrary."[3] In any case, there can be no disputing the fact that, on the whole, the word designating a concept has fewer connections with it than does the concept with its definition and its contents. Granted that verbal signs are sometimes "motivated" (to use Saussure's expression) and that there is occasionally a resemblance between the symbol and what it symbolizes; granted too that, as Benveniste reminds us, to the speaker the word does not seem arbitrary at all (young children think of the names of things as physically a part of them—a mountain has always had its name, even before men, by looking at it, find out what that name is!), still, the conventional character of the verbal sign is incontestable, as the multiplicity of languages proves. Note that "conventional" does not mean simply "arbitrary"—verbal signs depend upon implicit or explicit agreements based on custom, in contrast to symbols,

[2] The term "value" is for Saussure a technical one; see, for example, pp. 115ff. and 79ff. in *Course in General Linguistics*. This has a bearing on Piaget's argument in section 18. [Trans.]

[3] *Course in General Linguistics*, pp. 131ff. [Trans.]

which may be of individual origin, as in symbolic games or dreams.[4]

It follows from what we have just said that the relations between synchronics and diachronics must be different in linguistics than in other domains, where structure belongs, not to the means of expression, but to the expressed, to the signified rather than to the signifier, in short, to realities which have intrinsic value and normative power. The defining character of norms is that they are obligatory, that they conserve their own value by binding men to such conservation. *Their* equilibrium at any given time depends upon their antecedent history, for the distinctive character of development here is that it is always directed toward such equilibrium.[5] The history or rather chronicle of a word, however, may simply consist of a series of changes of meaning without any mutual relations except such as result from the necessity of answering to the expressive requirements of the successive synchronic systems to which the word belongs. Normative and conventional structures are, therefore, at opposite poles as regards the relations between synchronics and diachronics. As for value structures, such as those considered in economics, they occupy an intermediate posi-

[4] See Note 8, Chapter VI, and compare *Origins of Intelligence in Children,* pp. 189ff., where Piaget speaks of the symbol and the sign as "the two poles, individual and social, of the same elaboration of meanings." [Trans.]

[5] In the case of norms this equilibrium depends on the possibility of ever more dramatic reversals, while in linguistics it is rather a question of mere oppositions, without ruling out a mechanism (as yet, however, very little understood) of collective self-regulation.

tion; consideration of the development of the means of
production ties it to diachronics while the study of the
interaction of economic values ties it to synchronics.[6]

---

[6] Whereas the essentially descriptive and taxonomic linguistics
of the "distributionalists" (Bloomfield and his collaborators) did
not advance beyond the Saussurian synchronic structuralism,
Trubetzkoy, by applying the principle in terms of which Saussure had defined the systematic character of language, namely
the principle of opposition, to *sounds,* became the founder of
phonology (previously it had chiefly been applied to *signifiers*).
The phonological system of a language is, like its system of
signifiers, a *structure,* a network of *differences.* Roman Jakobson's theory of "distinctive features" is a further refinement of
this phonological structuralism. The "glossematicians" (Brondal,
Hjelmslev, Togeby), not to mention J. Trier and his *"Bedeutungs-
systeme,"* are likewise "structuralists," albeit of a somewhat
different stripe: for Hjelmslev a structure is "an autonomous
system of internal dependencies." "For any given process there
is a hidden system," process itself being the transition from one
to another system. Yet this transition is not regarded as *formative.*
Hjelmslev's somewhat esoteric vocabulary makes it hard to
summarize his views, but it is worth noting that he too endorses
the hypothesis of a "sublogic" which functions as common
source of both logic and language. Since he tends to emphasize
"dependencies" rather than "transformations," however, Hjelmslev's linguistic structuralism remains static. [The reader who
has some difficulty finding his way through this thicket of names
and technical terms may want to turn to Milka Ivić's *Trends in
Linguistics* (translated from the Serbo-Croat by Muriel Heppell,
Mouton, 1965). Ivić makes some of the points Piaget presumably
has in mind here: for example, p. 142, a propos of Jakobson:
"Synchronic investigation should be of primary interest . . . but
that does not mean that history of language should be excluded. . . . History of language acquires its real sense if the
evolution of a language is seen as the evolution of the system
as a whole. . . ." p. 146: "He (Jakobson) was the first to approach the history of language with the aim of revealing the
inner (linguistic) logic of language evolution." According to
Ivić, Hjelmslev was strongly influenced by Carnap.]

15. *Transformational Structuralism and the Relations between Ontogenesis and Phylogenesis*

It is extremely interesting that, despite the very strong arguments for keeping linguistic structuralism within synchronic confines, present-day linguistic structuralism, as represented by the work of Zellig S. Harris, and, above all, his pupil Noam Chomsky, has, as regards syntax, a clearly "generative" orientation. This interest in "generation" is, in Chomsky's work, accompanied by an attempt to formalize linguistic transformation, as indeed it should be. (It is to be noted that the transformation rules have a certain regulative power as well: they "filter out" certain structures as "ill formed.") Chomsky's theories place linguistic structures among those maximally general structures which derive their wholeness, not from descriptive and static laws, but from laws of transformation; and whose orderliness is a matter of self-regulation.

The reasons for this striking change of perspective are of two kinds, well worth analyzing, because they are relevant not only to the comparative study of *structures* but also to the comparative study of *theories of structure*. Their effect is, therefore, truly interdisciplinary.

The first relates to the recognition of what Chomsky has called the "creative" aspect of language, earlier noted by Harris and M. Halle, which comes to light mainly in individual acts of speech (as opposed to language in a more abstract sense) and is therefore studied by psycholinguistics. In fact, after decades of suspicion toward psychology, linguists have re-established their connection

with it through psycholinguistics.[7] Chomsky is very much
involved in these new developments.

> A central topic of much current research is what we may
> call the creative aspect of language use, that is, its un-
> boundedness and freedom from stimulus control. The
> speaker-hearer whose normal use of language is "creative"
> in this sense must have internalized a system of rules that
> determines the semantic interpretations of an unbounded
> set of sentences; he must, in other words, be in control of
> what is now often called a *generative grammar* of his
> language.[8]

The second reason for Chomsky's interest in the transformation laws of "generative grammar" is quite para-
doxical, since it seems at first sight oriented toward a
radical "fixism" opposed to any notion of genesis and
transformation: the idea that grammar has its roots in
reason, and in an "innate" reason. In a comparatively
recent work, *Cartesian Linguistics,* Chomsky goes so far
as to proclaim Arnauld and Lancelot, and Descartes, his
ancestors—the former on account of their *Grammaire
générale et raisonnée,* the latter on account of his analysis
of the connectedness of language and *"esprit."* His own
theory, according to which the rules of transformation
whereby "derived sentences" are formed yield these *from*

[7] See, for example, Sol Saporta and Jarvis R. Bastin, *Psycholinguistics: A Book of Readings* (New York: Holt, Rinehart &
Winston, 1961). [Trans.]

[8] "Persistent Topics in Linguistic Theory," *Diogenes,* No. 15
(Fall 1965), p. 13. [*Diogenes* is published in English as well as
in French. The translator thought it made better sense to turn
directly to the English version instead of translating the French.
But there seems to be quite a gap between the two.]

certain fixed "kernel sentences" of subject-predicate form (the connecting link between language and logic) is, in his opinion, incipient in the writings of these "Cartesian linguists." That this new rationalism, which Chomsky describes as "a return to traditional ideas and viewpoints rather than a radical innovation in linguistics or psychology,"[9] completely inverts logical positivism does not bother him in the least. While the logical positivists, enthusiastically followed by Bloomfield, wanted to reduce mathematics and logic to linguistics and the entire life of the mind to speech, Chomsky and his followers base grammar on logic and language on the life of reason.

This deliberate inversion is just as clear on the terrain of methodology. Emmon Bach, in a fascinating article[10] which, for all its courtesy and fairness, is a severe critique of logical positivism and the linguistic methods it inspired, provides a penetrating analysis of the epistemological presuppositions of Chomsky's structuralism. According to Bach, the remarkable work of American linguists between 1925 and 1957 was altogether Baconian in method: inductive data gathering, heterogeneous domains of research—phonetics, syntax, and so on—pyramidally arranged and more or less loosely connected in retrospect, distrust of "hypothesis," indeed of ideas, a program of making "protocol sentences" serve as epistemological "bases," and so forth. Chomsky's method, on the other hand, which Bach contrasts with the Baconian by placing it under the auspices of Kepler, turns on the recognition that there *are* no such "bases," that sci-

[9] Chomsky, *op. cit.*, p. 20.
[10] "Structural Linguistics and the Philosophy of Science," *Diogenes*, No. 15 (Fall 1965), pp. 111–127.

ence calls, instead, for "hypotheses" (indeed, for those "least probable" hypotheses which Karl Popper[11] could call "best" because, when falsifiable, they enable us to eliminate so large a number of consequences at one stroke). Instead of looking for an inductive step-by-step procedure to help us collect the properties of particular languages and ultimately language in general, Chomsky inquires: What grammatical postulates are necessary and sufficient to describe the universal principles of language structure and to furnish a general method for selecting a grammar for any given particular language? Chomsky actually arrived at this conception of linguistic structure by combining mathematico-logical concepts and techniques of formalization (algorithms, recursive devices, abstract calculi, and especially the algebraic concept of the monoid or semigroup) with ideas taken from general linguistics on the one hand (especially the conception of syntax as "creative") and from psycholinguistics on the other (for example, the idea of the speaker-hearer's "competence" in his own language).

Briefly put, his theory is as follows: Employing recursive methods, we can obtain a set of rewriting rules of the form $A \rightarrow Z$, where A is a symbol for categories like "noun phrase," "verb phrase," and so on, and Z is a terminal or non-terminal symbol string. We obtain derived sentences by applying transformation rules to non-terminal strings (strings which can be further rewritten) and it is the set of transformation rules which constitutes any particular "generative grammar," that is, a grammar

[11] See *The Logic of Scientific Discovery* (New York: Basic Books, Inc., 1959). [Trans.]

"capable of establishing connections between semantemes and phonemes in an infinity of possible combinations."[12]

This genuinely structuralist procedure of devising a coherent system of transformations (in effect, more or less complex "networks") is not only an excellent instrument for comparative study but possesses the additional interest of applying to "individual competence" (being the "internalized grammar" of the speaker-hearer) as well as to language as a social institution. For example, a number of psycholinguists—S. Ervin in collaboration with W. Miller and R. Brown together with U. Bellugi— have been able to use Chomsky's methods to reconstitute children's grammar: these turn out to be quite original and different from the grammar of adults. Such genetic applications of Chomsky's structuralism are noteworthy for several reasons. First, they do much to attenuate the contrast between "language" as a social institution and "speech" as individual performance (an opposition originally established by Dwight Whitney, taken over by Durkheim and Saussure, and since then more or less taken for granted) and to cast doubt upon the notion that the development of speech and with it of all individual thought consists merely in an adaptation to the collective norms.[13] Second, these linguistic studies of ontogenesis

[12] The phrase in quotes occurs on p. 21 of the French edition of the article in *Diogenes* cited above. [Piaget's account of *Syntactic Structures* is so excessively abbreviated as to obscure the distinction between "phrase structure rules," "transformation rules," and "morphophonemic rules."—Trans.]

[13] Would even the most civilized languages be the same as they are now if adults on the average reached an age of, say, 300 years and the gap between generations were appreciably larger than it is now?

in its reciprocal interaction with phylogenesis or social development are in line with similar tendencies in other disciplines—in biology as understood by Waddington and—if the reference be permitted—in genetic epistemology in its many aspects.

This idea of an interaction between ontogenesis and linguistic structure can today be met with even in areas where it would have been unimaginable before, namely, in the affective domain and that of unconscious symbolization. Bally long ago tried to develop a theory of "emotive speech" (*langage affectif*), speech which strengthens the natural expressiveness of everyday language; but Bally's "stylistic" showed primarily that this "emotive" speech tends to disintegrate the normal structures of language. Why not rather consider the hypothesis that feeling has a language of its own? Under the influence of Bleuler and Jung, even Freud finally came around to this view, though he had previously attempted to explain symbolization as a mechanism of disguise. Jung's theory of symbols as hereditary "archetypes" Freud rightly rejected, seeking their origin instead in ontogenesis. It seems that we are here on a terrain which, though not directly connected with linguistics, has nevertheless an important bearing on semiotics and on what Saussure called "general semiology."[14] Lacan's *Ecrits* should also be mentioned in this connection: All psychoanalysis, he points out, has speech for its medium; that of the analyst, who normally says very little, but chiefly the speech of the patient; indeed, the psychoanalytic process consists essentially in the patient's "translating" his unconscious

---

[14] See *Course in General Linguistics*, p. 16. [Trans.]

individual symbols into a conscious and public language. Taking off from this new idea, Lacan has tried to use linguistic structuralism and familiar mathematical models to devise new transformation rules which would make it possible for the irrational ingredients of the unconscious and the ineffable features of private symbols to make their entry into a language really designed to express the communicable. No matter what the outcome, the project in itself is of great interest. But not until the "uninitiated" have "clarified" Lacan's results will we be in a position to gauge their value. (Psychoanalysts reserve the epithet "initiated" for members of the local psychoanalytic chapter. The point we are making is that though to appraise a theory one must obviously be in possession of the relevant facts, be initiated in that sense, not until a theory has become detached from the milieu in which it originated can it attain the rank of possible truth.)[15]

16. *Are Linguistic Structures Social Formations, Innate, or the Results of Equilibration?*

The fascinating mix of geneticism and Cartesianism characteristic of Chomsky (*Le mélange si intéressant de génétisme et de cartésianisme qui caractérise Chomsky*)

---

[15] See p. 139 below, where the same word, "de-centering," which here crops up to make the point that "une vérité n'est accessible comme telle qu'une fois *décentrée* des influences qui lui ont donné naissance," is used to refer to the individual subject's "work" of overcoming his spontaneous intellectual egocentricity. [Trans.]

has led the latter to defend a thesis which, quite unexpectedly, ties "innate ideas" to that heredity which, according to certain biologists, will eventually account for nearly all mental life:

> . . . if the grammars of natural languages are not only intricate and abstract, but also very restricted in their variety, particularly at deeper levels, it becomes necessary to challenge the widespread assumption that these systems are "learned" in some significant sense of this term. It is perfectly possible that a particular grammar is acquired by differentiation of a *fixed innate scheme,* rather than by slow growth of new items, patterns, or associations . . . and the little that is known about the structure of language suggests that the rationalist hypothesis is likely to prove productive and fundamentally correct in its general outline.[16]

Here we have the hypothesis latent in the work of those authors whose structuralist tendencies lead them to distrust psychogenesis and history but who, for all that, do not want to turn structures into transcendent essences. Chomsky's position is rather more subtle in that, according to him, the transformation processes by which particular grammars come to be differentiated go into action in the course of development: only the "kernel," the "fixed scheme," is innate, along with the most general features of the transformational structure, while the variety of natural language derives from the "creative" aspect of language mentioned earlier. By distinguishing between "kernel" and "husk," "deep" and "surface"

[16] Chomsky, *op. cit.,* p. 19.

structures, Chomsky allows for both description and formalization. Nevertheless, he leaves us with the fundamental problem of the nature and origin of the "fixed innate schema."

There is, first of all, the biological question—even when a trait is recognized as hereditary, the question of its formation remains. It is hard enough to understand how the cortical centers for language developed in the course of "hominization." As we said earlier, mutation and natural selection are generally inadequate solutions; and, since speech is born of communication among individuals so that its development seems to presuppose its existence, they are all the less helpful here. If, furthermore, the genes responsible for language must transmit not only the capacity for learning a language, that is, the ability to acquire it "from outside," but a fixed innate scheme that forms language from *within*, the problem becomes so complex as to seem beyond solution. And if, above and beyond this, the "formative kernel" is endowed with reason, so that reason itself must be regarded as hereditary, there seem to be only two possible responses: preformationism (but why, then, must we wait for man for speech to come upon the scene when chimpanzees and bees are already so "sympathetic"?) or some version of Waddington's theory of interaction between genotype and environment (see pages 49–50 above).

Now when we descend to the terrain of ontogenesis, where the details of acquisition and transformation are verifiable, we meet with facts which, though presenting certain connections with Chomsky's assumption, differ as to the importance or range of the hereditary points of

departure (see Sections 12 and 13). And the reason for
this is, undoubtedly, that where Chomsky sees only two
alternatives—either an innate schema that governs with
necessity, or acquisition from outside (cultural and therefore variable determination such as cannot account for
the limited and necessary character of the schema in
question)—there are in fact three possibilities. There is
heredity versus acquisition from outside, true; but there
is also the process of internal equilibration. Now such
equilibration processes, such self-regulation, also yield
necessities; we might even say that their results are more
necessary than those determined by heredity, for
heredity varies much more than do the general laws of
organization by which the self-regulation of behavior is
governed. Moreover, heredity applies only to traits that
either are or are not transmitted, bears on "contents";
whereas self-regulation sets a direction compatible with
a construction that becomes necessary precisely in being
directed.

There are at least two reasons for trying the equilibration hypothesis and suspending Chomsky's innatism while
preserving the rest of his theory. First, the equilibration
hypothesis allows for the construction of cybernetic
models of linguistic structures such as S. Saumjan of the
Moscow Academy of Science has been trying to devise.[17]

[17] See Saumjan's "Cybernetics and Language," *Diogenes,* No.
15 (Fall 1965), especially pp. 142ff., where he contrasts his
own generative grammar with Chomsky's in the following terms:
"In the applied generation model the primary notion is that of
the *field* of transformation and not . . . transformation itself.
This field is intended as a system of special operators which I
call 'relators' . . . the concept of transformation . . . represents
only an element of the field of transformation. . . . Just as

Even if this project should prove unrealizable, this fact would of itself be highly instructive for, if it is true that, as Bar Hillel has suggested,[18] formal grammatical systems do not furnish decision procedures, the limits of formalization would apply here as well; here too it would be necessary to drop the notion of a "basis" that contains everything in advance, and layer by layer *construction* would have to take the place of *axiomatization*.

Second, the relatively late appearance of language in the course of the second year of life seems to confirm the constructivist thesis. For why should speech begin at this level of development and not earlier? Contrary to the too facile explanations by conditioning, which imply that language acquisition starts as early as the second month, the acquisition of language presupposes the prior formation of sensori-motor intelligence, which goes to justify Chomsky's ideas concerning the necessity of a prelinguistic substrate akin to rationality. But this intelligence which antedates speech is very far from preformed from the beginning; we can see it grow step by step out of the gradual coordination of assimilation schemes. It has therefore occurred to H. Sinclair, to whose work we shall turn in a minute, that the processes of repetition, ordering, and associative connecting whereby the sensori-motor schemata become coordinated *themselves* contain the source of Chomsky's "monoid." If her hypothesis proves

---

phonology was transformed from the theory of phonemes to the theory of phonological oppositions . . . so must transformation grammar change from the theory of transformations to that of transformation fields. . . ."

[18] "Decision Procedure in Natural Language," *Logique et Analyse,* 1959.

warranted, we would have an explanation of linguistic structures that dispenses with too heavy-handed an innatism.

17. *Linguistic and Logical Structures*

Let us return to the problem with which we started, which remains one of the most controversial issues of structuralism and, indeed, of epistemology in general. Any serious answer to the question of how linguistic and logical structures are related must, of course, be provisional. Even a Soviet linguist like Saumjan, working in a culture center where only a few years ago the Pavlovian theory of language as a "secondary signal system" seemed to have taken care of all problems, admits that the question of the relation between language and thought is "one of the hardest and most profound of contemporary philosophical problems." We can obviously not even begin to solve the problem here; all we mean to do is to indicate what, from the structuralist perspective and taking recent developments in linguistics into account, the state of the question is.

But we must backtrack and recall two important facts: The first is that, since Saussure and many others, we know that verbal signs exhibit only one aspect of the semiotic function and that linguistics is really only a limited though especially important segment of that more inclusive discipline which Saussure wanted to establish under the name of "general semiology." The symbolic or semiotic function comprises, besides language, all forms of imitation:[19] mimicking, symbolic play, mental imaging, and so on. Too often it is forgotten that the development of representation and thought (we are not as yet speaking of properly logical structures) is tied to this *general* semiotic function and not just to language. How otherwise could we explain that deaf-mute children (those, that is, whose brain has not been damaged) play at make believe, invent symbolic games and a language of gestures? By studying their concrete logical operations —seriation, classification, conservation, and so forth—as investigators like P. Oleron, H. Furth, M. Vincent, and F. Affolter[20] have done, one can watch the development of these logical structures; this development is occasionally slowed down, but much less so than in the case of children blind from birth (studied by Y. Hatwell). In the case of blind children, language, which is quite normal, only slowly makes up for gaps in their sensori-motor schemata, whereas the deaf-mute child's deprivation of language does not interfere with the development of operational structures (the one-to-two-year average retardation as compared to normal children being attributable to lack of social stimulation).

The second fact to be recalled is that intelligence precedes language not only ontogenetically, as we saw in Section 16 and as our remarks about deaf-mute children confirm, but also phylogenetically—the numerous studies of the intelligence of the great apes amply prove this.

---

[19] See Piaget's *Play, Dreams and Imitation in Childhood* (New York: Norton, 1951). [Trans.]

[20] On account of the ingenuity of its techniques and the copiousness of its illustrations, Furth's *Thought without Language* is particularly interesting.

Even sensori-motor intelligence already involves certain
definite structures (order, subordination schemes, correspondences, and so on), which derive from the activity
of coordinating acts and are prior to rather than derivative from language.

This much said, it is obvious that, if speech depends
upon an at least partially structured intelligence, the
reverse is also true; speech structures this intelligence, and
here begin the real problems. They have certainly not
been solved. But the two methods now at our disposal—
transformational analysis (see, for example, M. D. S.
Braine's psycholinguistic studies of the acquisition of
syntax) and *operational analysis* (see Inhelder's, Sinclair's, and Bovet's experimental studies of the acquisition
of logical structures)—enable us to analyze the correlations between syntactic and operational structures, at
least at certain particular points; we are even in a position
to guess just where there is interaction between the two
and which of the linguistic or logical structures are prior,
which posterior, in the process of construction.

Let us, for example, briefly consider H. Sinclair de
Zwaart's novel and precise experiments.[21] She formed
two groups of children, choosing as criterion for their
"operational level" their ability or inability to deduce the
conservation of a certain volume of liquid upon seeing
it poured into containers of different shapes. The first,
clearly "pre-operational," group was made up of children
who denied conservation, while those of the second group
admitted it at once and explained it in terms of prin-

---

[21] H. Sinclair de Zwaart, *Acquisition du langage et développement de la pensée* (Paris: Dunod, 1967).

ciples of reversibility and compensation. She also analyzed the speech of these children by asking them questions that did not refer to the conservation experiments but rather to couples or collections of things which they were to compare with one another—a long and a short pencil, a long thin one and a short thicker one, a collection of four or five blocks and a collection of just two, and so on. Next she asked the children to carry out certain orders: "Give me a smaller pencil" or "Give me one that is smaller and thinner," and so forth.

She found that there is a systematic difference in the language of the two groups. The children in the "pre-operational" group rarely use any except scalar adjectives: "That one is big"; "This one is little"; or "Over there there's a lot"; "Here there aren't a lot"; and so on. The children in the second group, on the other hand, employ vector vocabulary: "That one is bigger than the other"; "There are more here"; and so on. Besides, where the things they are asked to compare differ in more than one respect, the children in the first group tend to ignore one or else juxtapose "kernel" sentences: "That one is big, this one is little; that one (the first) is thin, this one is thick," and so forth. Those in the second group, on the other hand, note binary connections like "That one is longer and thinner, the other shorter and thicker." Obviously, there is a correlation between the operational and the linguistic level, and we see immediately how the verbal structuration of the children in the second group can help their reasoning. The children in the first group do understand the expressions which they do not as yet themselves employ, as is shown by their ability to execute orders given in terms of that higher echelon vocabulary.

Mrs. Sinclair therefore subjected them to linguistic training, difficult, but possible. After this training, she reexamined them and found their progress quite small; about one in six now recognized that the quantity of liquid remained the same.

There must, of course, be additional experiments of this sort. It seems that on the level of "concrete operations" (see Section 12) operational structure precedes linguistic structure, the latter somehow growing out of the former to rely upon it subsequently. It remains to be investigated by some analogous procedure what exactly happens at the level of "propositional operations," where the language of children is modified so strikingly while their reasoning becomes hypothetico-deductive. If we can say today that it is as much as demonstrated that language is not the source of logic, that Chomsky is right in grounding language in reason, it must nevertheless also be said that the detailed study of their interaction has only begun. What is wanted is further experimentation and correlative formalization.

VI

STRUCTURAL ANALYSIS IN THE SOCIAL SCIENCES

18. *"Analytic" versus "Global" Structuralism*

By the definition of structure proposed in our opening chapter, *all* the social sciences yield structuralist theories since, however different they may be, they are all concerned with social groups and subgroups, that is, with self-regulating transformational totalities. A social group is evidently a whole; being dynamic, it is the seat of transformations; and since one of the basic facts about such groups is that they impose all sorts of constraints and norms (rules), they are self-regulating.

But there are at least two important differences between this sort of "global" structuralism and the deliberate analytic structuralism of, say, Lévi-Strauss. First, where the former speaks of "emergence," the latter speaks of "laws of composition": Durkheim's structuralism, for example, is merely global, because he treats totality as a primary concept explanatory as such; the social whole arises of itself from the union of components; it "emerges." His collaborator Marcel Mauss, on the other hand, is regarded by Lévi-Strauss as the originator of authentic anthropological structuralism because, especially in his studies of the gift, he sought and found the details of transformational interactions. Second, whereas "global" structuralism holds to systems of observable relations and interactions, which are regarded as sufficient unto themselves, the peculiarity of authentic (analytic) structuralism is that it seeks to explain such empirical systems by postulating "deep" structures from which the former are in some manner derivable. Since structures in this sense of the word are ultimately logico-mathematical models of the observed social relations, they do not themselves belong to the realm of "fact." This means, among other things, as Lévi-Strauss points out repeatedly, that the individual members of the group under study are unaware of the structural model in terms of which the anthropologist interprets constellations of social relations.

The two characteristics of analytic structuralism are obviously connected; in fact, the search for "deep" structures is a direct consequence of the interest in the details of transformation laws. Once this is recognized, certain striking analogies between structuralism in anthropology, in physics, and in psychology become apparent. The social structure, like causality in physics, is a theoretic construct, not an empirical given. It is related to the observable social relations as, in physics, causality is related to physical laws, or as, in psychology, psychological structures do not belong to consciousness but to behavior (only when there is some sort of dis-adaptation does the individual become aware of structures, and this awareness is always quite dim and partial).

Let us start out by considering sociology and social psychology, two disciplines whose boundaries are becoming increasingly fluid (as can be said of all demarcations that depend more upon a desire for professional autonomy than upon the nature of things). Kurt Lewin's work will illustrate what hopes and partial successes are characteristic of this form of structuralism and also its necessarily interdisciplinary character.

A pupil of Wolfgang Köhler in Berlin, Lewin very early conceived of the idea of applying Gestalt structures to social relations and for this purpose generalized the notion of "field." While the perceptual or more generally the cognitive field is for the Gestaltists simply the ensemble of simultaneously apprehended elements, Lewin, to analyze affective and social relations, proposed a notion of "total field" encompassing the individual subject with all his needs and dispositions. The logical complexity of this notion of "total field" derives from the fact that disposition concepts generally, and in particular the concept of need, cannot be defined apart from reference to an environment and that neither a physical, nor even a purely biological, but only a psycho-biological account of environment will be sufficiently concrete to enable us to predict or understand the behavior of the human subject. Thus, the mere physical presence of an object will not determine behavior. In the first place, only as having a "demand value" (*Aufforderungscharakter*) or "valence" does it enter into the dynamics of the situation. In the second place, its accessibility, which depends not only upon proximity or distance, but also upon the presence or absence of "barriers" (many of which are psychological, such as prohibitions and inhibitions of various kinds), must be taken into account. Since the "demand value" of environmental objects and the "needs" of the individual are correlative, only some sort of field conception will be adequate to Lewin's purposes, and since psychodynamic locomotion depends to so large an extent upon possible kinds of connections between "regions" (which may intersect, be separated from one another, or contain one another) it is no wonder that Lewin attempts to analyze his total field in topological terms. Unfortunately, the psychological topology is not really mathematical, that is, there is not one known topological theorem that can be given a direct psychodynamic interpretation; nevertheless, the basic topological conception of a purely qualitative analysis of spatial relations does find application. Lewin offers his topological analysis as determining *possible* "movements" or "paths." To account for the subject's *actual* behavior the concept of force, hence of "vectors," must be introduced. This utilization of physical concepts has the great advantage of making field diagrams applicable to dynamic psychology and of suggesting network structures.[1]

It is by means of such purely structuralist methods that

[1] The foregoing paragraph is a slightly expanded rendering rather than a direct translation of Piaget's text. See Kurt Lewin, *A Dynamic Theory of Personality* (New York: McGraw-Hill, 1935). [Trans.]

Lewin and his pupils (Lippit and White in the United States; Dembo, Hoppe, and particularly Zeigernik in Berlin) have constructed a social and affective psychology which has seen great development in the United States and is one of the principal sources of the numerous current studies of "group dynamics." (A special Institute for the Study of Group Dynamics, headed by Cartwright, now exists at Ann Arbor.) The latter, which have proliferated into great variety, furnish a handsome example of analytic structuralism. On the observational level, the rule that all the facts deserve meticulous description is strictly adhered to, but causal explanations are sought by having recourse to structural models; there are even some specialists in the mathematizing of structural models of small groups—R. D. Luce in the United States and Cl. Flament in France.

We need say little here about microsociology and sociometrics, since these retain, to the extent that they are at all concerned with social structure, the global orientation we described earlier, so that, even when the observed constellations of social relations are multiplied and imbedded in a "dialectical pluralism," no genuinely structuralist theory is forthcoming; or they continue to adhere to the statistical methods whereby relations are, no doubt, quantified, but not in any sense explained.

Macrosociology, on the other hand, obviously brings up all the major structural problems. We shall reserve a discussion of Althusser's "translation" of Marxism into structuralism for Chapter VII, since this will enable us to take up the interesting problem of dialectic in general. Here we wish rather to consider the work of Talcott Parsons, whose "structural-functional" method brings us back again to the question of the relation of structure and function.

In the Anglo-Saxon countries, the concept of structure tends to be reserved for observable relations and inter-actions. Parsons' work deserves special mention partly because it goes beyond this too modest empiricism. Thus, his definition of structure as a stable disposition of the elements of a social system impervious to externally im-posed disturbances has led him to develop a theory of social equilibrium sufficiently worked out to make him say to one of his collaborators that he would like to see it formalized. As for functions, Parsons conceives them as entering whenever structures, in the just defined sense, become adapted to new situations.

For Parsons, structure and function together belong, accordingly, to a "total system" that is conserved by "regulations," and the problem that has chiefly occupied him is to explain how individuals come to integrate com-munal values. This is where his theory of "social action" comes in, which analyzes the various alternative courses of action open to the individual in terms of whether or not he submits to the collective values.

The work of M. J. Levy, who reduces structures to observable uniformities, and functions to diachronic manifestations of structure, goes back to Parsons'. But in our opinion *norms, values* (whether spontaneous or normative), and *symbols* in the wider sense, or signs (see Section 14) call for somewhat different analyses of the relations between synchronics and diachronics. On the other hand, Parsons' way of linking functions to values should be taken very seriously. In the social context, structures, no matter how "unconscious," express themselves sooner or later in the form of norms or rules to which individuals are, to a greater or lesser extent, subject. Now, however convinced one may be of the permanence of structures themselves (see Section 19), the *rules* generated by them can nevertheless change their function, as is shown by changes of value; *values* do not, of themselves, have "structure," except precisely to the extent that certain forms of value, such as moral value, are based on *norms*. Value seems, then, to point up a distinct dimension, the dimension of function; thus, the duality and re-established interdependence of value and norm seem to testify to the necessity of distinguishing and connecting structure and function.

It is this problem of structure and function which has dominated structuralist theories of economics. When F. Perroux defines structure in terms of "the proportions and ratios characterizing an economic ensemble localized in space and time," the very narrowness of this definition shows up the difference between economic structure and structure as we have spoken of it until now. But this difference is not due to the fact that economists limit the notion of structure to observable relations. Jan Tinbergen, for example, thinks of structuralist economics as "the study of those not directly observable properties which concern the way in which the economy reacts to certain changes"; in econometrics these properties are expressed in terms of coefficients "and the ensemble of such coefficients contains two kinds of information": first, it provides a blueprint of the economy; second, it indicates the paths of reaction to certain modifications. There could be no clearer indication that the economic structure is a functioning one, since it is capable of "reacting"; it cannot be thought of apart from its functions.

As for the nature of this structure, emphasis was at first laid on the analysis of states of equilibrium, but once the dynamics of business cycles became the chief problem for economic theory, it became necessary to modify this view by allowing precisely for the notion of operation or function. For Marshall, the solution lay in extending the concept of equilibrium structure by introducing, as in physics, "displacements of equilibrium." Keynes, on the other hand, tried to account for cycles in terms of the predictions and calculations of the economic agent. But for both, and not only for them, the structural concept of equilibrium became, as G. G. Granger has pointed out, an "operator" with which to account for business cycles.

But it is not only this primacy of functioning which distinguishes economic structures; their essentially probabilistic character (which is of course connected with the primacy of activity) is just as important: Economic equilibration or self-regulation is not strictly operational but of the feedback type. This holds not only for the individual economic agent but for the econometrist's larger economic ensembles as well. It has been said, by G. G. Granger, that game theory eliminates the need for a consideration of psychological factors from economics, and this is no doubt true so long as one is thinking of the rather abbreviated psychology of Pareto or Böhm-Bawerk. But once the role of "strategies" in behavior in general is taken into account (we are speaking of behavior, not of consciousness), once it is recognized that game theory is applicable to affects, to perception, and to cognitive development, economic structures become, via game theory, very closely linked to the affective and cognitive regulations of the *subject*.

That feedback systems play an important role in econometrics and macroeconomics is so well known a fact that we need only mention it here.

It remains to say something, however abbreviated, about the structure of law. In contrast to spontaneous *values*, social structures concerned with *norms* exhibit a strikingly "operational" character. Thus, as is well known, Hans Kelsen described the juridic structure as a pyramid of norms held together by a general relation of implication which he called "imputation": At the top of the pyramid there is the "fundamental norm," the source of the legitimacy of the whole and, in particular, of the constitution. Laws and the power of the courts derive their legitimacy from the constitution. Acts of government are immediately justified by the law, mediately by the constitution. This "implied" legitimacy extends all the way down to "arrests" (whose legal character depends on it) and to the numerous "individualized norms" (indictments, elections, diplomas, and so on). Kelsen's pyramid can easily be given the form of an algebraic network (each norm being the "application" of higher norms and the "creation" of inferior ones, except, of course, for the "fundamental" and the "individualized" norms), but that does not answer the question as to its *nature*. The sociologist will answer that it is a *social* entity. But Kelsen replies that the *ought* is not reducible to the *is; norms* are not *facts*. The juridic structure is intrinsically and irreducibly normative. So long as we remain with the derived norms this answer does very well. But what of the "fundamental norm"? If its legitimacy does not *consist* in its being "acknowledged" by those subject to the law, on what, then, does it depend? On human nature? That is the answer of the natural right theorists, an answer no doubt satisfactory to those who believe in the permanence of human nature, but one which cannot but seem circular to those who want to understand this human nature in terms of its formation.

19. *The Anthropological Structuralism of Claude Lévi-Strauss*

The brevity of the preceding remarks can be remedied to some extent by dwelling somewhat longer on anthropology, the "synoptic" social science in that, being chiefly concerned with "elementary" societies, it necessarily studies psycho-social processes in their connection with linguistic, economic, and legal structures. This more detailed discussion of structural anthropology is further justified by the fact that Lévi-Strauss, its most distinguished representative, is the very incarnation of the structuralist faith in the permanence of human nature and the unity of reason. His structural models—neither functional, nor genetic, nor historical, but deductive—are in some manner paradigmatic; they show what could be achieved in the social sciences by employing structuralist methods. Furthermore, we cannot help thinking that our own constructivist theory of cognitive structure (see Sections 12 and 13) is intimately connected with Lévi-Strauss' doctrine of the primacy of structure in social life.

To perceive the novelty of his method, we suggest that

the reader turn to *Le totemisme aujourd'hui* (translated into English by Rodney Needham and published under the title *Totemism*),[2] where it is applied to what was for long the key concept of ethnology, namely the pseudo-entity "totemism." We cannot here linger over the fascinating descriptive details. Suffice it to highlight what seems to us the fundamental principle of Lévi-Strauss' structuralism—the thought that "all social life, however elementary, presupposes an intellectual activity in man of which the formal properties cannot, accordingly, be a reflection of the concrete organization of society" (English edition, p. 96). Durkheim, who, together with Mauss, had established the systematic and coherent character of "primitive" classifications such as are involved in "totemism," already admitted as much, as is shown by a remarkable passage in *Les formes élémentaires de la vie religieuse*, which Lévi-Strauss quotes at length (English edition, p. 96). But whereas Durkheim only too often insisted nevertheless on the primacy of the social over the intellectual, Lévi-Strauss recognizes that only by inverting this relation can Durkheim remain faithful to his own best thoughts. Behind the "concrete" social relations there is always "conceptual structure," unconscious, no doubt, and therefore discoverable only by elaborating abstract structural models, but nonetheless formative.

Anthropological structuralism is, accordingly, firmly synchronic, but the synchronic emphasis is placed somewhat differently here than in structural linguistics. In the first place, it is at least partly motivated by our irremediable ignorance of the first origins of beliefs and customs

[2] Boston: Beacon Press, 1963.

(see English edition, p. 70). In the second place, the system of beliefs and customs studied by the anthropologist is less subject to change than the language systems studied by the linguist, because "customs are given as external norms before they give rise to internal sentiments, and these norms, which are not feelings, determine the sentiments of individuals as well as the circumstances in which they may, or must, be displayed." (English edition, p. 70.) Now the norms themselves depend upon structures, which are permanent, so that this sort of synchronics is somehow an expression of an *invariant diachronics*. This does not mean, of course, that Lévi-Strauss would want to abolish history; only, the changes brought about by history do not affect the human mind itself and, furthermore, their analysis again requires recourse to "structures"—diachronic instead of synchronic.[3] History bears on the human mind only because

it proves indispensable for cataloguing the elements of any structure whatever, human or non-human, in their entirety. It is therefore far from being the case that the search for intelligibility comes to an end in history as though this were its terminus. Rather, it is history that serves as the point of departure in any quest for intelligibility . . . history leads to everything, but on condition that it be left behind.[4]

That this sort of position is "anti-functionalist" almost

[3] "*De facto* and *de jure*, there are diachronic as well as synchronic structures. . . ." R. Bastide, ed., *Sens et usages du terme structure* (Paris, 1962), p. 42.

[4] *The Savage Mind* (Chicago: University of Chicago Press, 1967), p. 262.

goes without saying; at least, Malinovski's kind of functionalism—"biological and psychological rather than strictly ethnological"—is severely criticized, turned down as "naturalistic, utilitarian, and affective" (*Totemism,* p. 56). When we consider how banal and at the same time dogmatic many of Malinovski's "Freudian" explanations are, it is no wonder that Lévi-Strauss should at times take a rather dim view of the explanatory import of biological and psychological considerations. And his penetrating critique of explanations in terms of affectivity ("the most obscure side of man"—*Totemism,* p. 69) which lose sight of the fact that "what is refractory to explanation can *ipso facto* not serve as explanation" (p. 69) deserves, of course, nothing but praise. Similarly, we are overjoyed to find him turning his back on the associationist psychology which is, unfortunately, still so much alive in certain quarters: ". . . it is the logic of oppositions and correlations, exclusions and inclusions, compatibilities and incompatibilities, which explains the laws of association, not the reverse. A renovated associationism would have to be based on a system of operations which would not be without similarity to Boolean algebra" (pp. 90ff.). But though this helps us to perceive that it is by a series of *logical* connections that *mental* relations become established (p. 80), and though we agree that what is wanted is "a reintegration of content with form" (p. 86) the problem of how eventually to coordinate sociological and anthropological structuralism with biological and psychological structuralism remains, it seems to us. And one thing is clear, in biology and psychology structural analysis must, at all levels, from homeostasis to operations, be supplemented by functional considerations.

To return now to the particular structural models of
Lévi-Strauss: while he took his departure originally from
linguistics, and while phonological or, more generally,
Saussurian, structures inspired his search for anthropological structures, the really decisive discovery for him
was, as is well known, that kinship systems are instances
of algebraic structures—networks, groups, and so on.
With the help of mathematicians like A. Weil and G. Th.
Guilbaud, he was able to give mathematical form to his
anthropological discoveries. And it turned out that not
only kinship systems, but all the "practices" and cognitive
products of the societies under study—the passage from
one system of classification to another, or from one myth
to another—lend themselves to this sort of structural
analysis.

Two important texts will help us grasp the meaning and
role of structures in anthropological explanation as
understood by Lévi-Strauss. The first is from the opening
chapter in *Structural Anthropology* (translated into
English by Claire Jacobson and Brooke Grundfest
Schoepf):

> In anthropology as in linguistics . . . it is not comparison
> that supports generalization, but the other way around. If,
> as we believe to be the case, the unconscious activity of the
> mind consists in imposing forms upon content, and if these
> forms are fundamentally the same for all minds—ancient
> or modern, primitive or civilized (as the study of the sym-
> bolic function, as expressed in language, so strikingly indi-
> cates)—it is necessary and sufficient to grasp the uncon-
> scious structure underlying each institution and each
> custom in order to obtain a principle of interpretation

valid for other institutions and other customs, provided, of course, that the analysis is carried far enough.[5]

This unchanging human mind, this "unconscious activity of the mind," is not to be confused with either Chomsky's "innate reason" or the "lived" (which is much rather to be "repudiated" so as subsequently to become reintegrated into an objective synthesis"—*Tristes Tropiques*, p. 50); it is a system of schemata intercalated between "infrastructures and superstructures":

> Marxism, if not Marx himself, has too commonly reasoned as though practices followed directly from *praxis*. Without questioning the undoubted primacy of infrastructures, I believe that there is always a mediator between *praxis* and practices, namely, the conceptual scheme by the operation of which matter and form, neither with any independent existence, are realized as structures, that is as entities which are both empirical and intelligible. It is to this theory of superstructures, scarcely touched on by Marx, that I hope to make a contribution. The development of the study of infrastructures proper is a task which must be left to history —with the aid of demography, technology, historical geography, and ethnography. It is not principally the ethnologists's concern, for ethnology is first of all psychology.[6]

This grand theory is saddled with one major problem, which is: once we have admitted the existence of structures as distinct from the system of observable relations

[5] *Structual Anthropology* (New York: Basic Books, 1963) p, 21.

[6] *The Savage Mind*, p. 130.

and interactions to which Radcliffe Brown (the English ethnographer who comes closest to structural analysis) continues to adhere, how are we to understand this "existence"? What does it consist in? Structures are not simply convenient theoretical constructs; they exist apart from the anthropologist, for they are the *source* of the relations he observes; a structure would lose all truth value if it did not have this direct connection with the facts. But neither are they transcendent essences, for Lévi-Strauss is not a phenomenologist and denies the primacy of the "me" or the "lived." The recurrent formula is that structures "emanate from the intellect," from the human mind as ever the same; this is why they are prior to, rather than, as Durkheim would have it, derivative from the social order; prior to the "mental" as well (whence the phrase quoted earlier, *"enchainements logiques unissant les rapports mentaux"*) and, *a fortiori*, to the "organic" (which, and rightly so, is held to explain affectivity but which cannot be the source of "structures"). But what manner of existence is left, then, for the mind, if it is neither social, nor mental in the subjective sense, nor organic?

By leaving the question unanswered and speaking simply of "natural structures" we risk a false analogy with doctrines of natural rights. So let us rather try to devise an answer ourselves. If it is, as Lévi-Strauss says, necessary to "reintegrate content with form," it is no less essential to recall that neither forms nor contents exist *per se:* in nature as in mathematics every form is content for "higher" forms and every content form of what it "contains." It is, of course, not sufficient, nor even strictly true, as we saw earlier, in Section 8, to say that "everything is structure"; *structure* must be defined more narrowly than *form*. How is this to be done?

Let us note first of all that, even though on the present theory everything can *become* structured, structures are, nevertheless, only one kind of "forms of forms," namely, such as are governed by the several extremely comprehensive limiting conditions we explained in our opening chapter: only self-regulating transformational systems are structures. So our question should run: how do forms acquire structural organization? When the structures in question are abstract logical or mathematical structures, we may say that the logician or mathematician "derives" them from "forms" by reflective abstraction. But there must also be a general formative process in *nature*, leading from forms to structures and establishing the self-regulation constitutive of the latter. In physics, biology, and psychology it is, as we saw in Sections 9, 10, 12 and 13 respectively, *equilibration* which accounts for the "selection" of the actual system from among the range of possibles; it is equilibration, again, which establishes homeostasis at its various organic levels and which explains the development of intelligence as well. May we not expect it to render similar services in the social sciences? Indeed, once it is recalled that every form of equilibrium is definable in terms of a "group" of "virtual" transformations and that a *state* of equilibrium must always be distinguished from the *process* of equilibration, the processes whereby equilibrium becomes established in these increasingly complex systems account not only for the regulations characteristic of each level but even for the form which these regulations take at the final stage, when they become "reversible operations." The equilibration of the "cognitive" and the "practical" functions contains all that is necessary for an explanation of the rational schemata: a system of lawful transformations and an opening to the possible, that is, the two conditions for transition from *temporal formation* to *non-temporal interconnection.*

From this perspective, there is no longer any need to choose between the primacy of the social or that of the intellect; the collective intellect is the social equilibrium resulting from the interplay of the operations that enter into all cooperation. Nor does intelligence precede mental life or the reverse; it is the equilibriated form of all cognitive functions. And the connections between intellect and organic life may be conceived of in the same way: though it would not do to say that all vital processes are "intelligent," it can be maintained that in the morphological transformations which d'Arcy Thompson studied more than a generation ago life is geometrizing;[7] today we may go so far as to say that in many respects life works like a cybernetic machine, an "artificial" or "general" intelligence.

But what, then, becomes of that unchanging human mind whose constancy Lévi-Strauss himself defends by appealing to the permanence of the "symbolic function"? We must admit that we do not really understand why the mind is more truly honored when turned into a collection of permanent schemata than when it is viewed as the as yet unfinished product of continual self-construction. *Must* the "symbolic function" be thought of as perma-

---

[7] See *On Growth and Form,* 2nd ed. (Cambridge, Mass.: Harvard University Press, 1942, 1952). This work, as well as his mineralogical studies, influenced Lévi-Strauss' earlier thought.

nent? Would it not be legitimate to think of what Saussure called the "sign" as having evolved from what he called the "symbol"?[8] Is not this what Rousseau meant in that passage concerning the primitive use of tropes which Lévi-Strauss cites with approval as correctly assigning a fundamental role to figurative language (*Totemism*, p. 102)? And when he speaks of metaphor as constituting a "first" or "primary" form of discursive thought, must we not take this to mean that there is something to follow after, or at least that there are "levels"? Granting that "thought untamed" (*la pensée sauvage*)[9] is always present among us, does it not nevertheless constitute a level of thought inferior to the scientific? Levels in a hierarchy imply stages of formation. It may well be asked, for example, whether the beautiful "primitive" classifications of which Lévi-Strauss speaks in *La Pensée sauvage*

---

[8] Saussure's trichotomy strikes us as more profound than Peirce's. Saussure differentiates the *index* (causally connected with what it signifies) from the *symbol* ("motivated") on the one hand and the *sign* ("arbitrary") on the other; the sign is necessarily social, because conventional, whereas the symbol may be, as in dreams, individual. Peirce's *symbol* is an approximate match for Saussure's *sign*, but it functions, not as a *subdivision* of "representational signifiers," but as a contrast to *icons* (roughly, images) on the one hand, *indices* (roughly the same as Saussure's) on the other. [In other words, the contrast between pre-representational and representational signifiers and between individual and social signifiers does not enter Peirce's classification at all.]

[9] See *The Savage Mind*, p. 219: ". . . Comte assigns this 'savage mind' to a period of history . . . while in this book it is neither the mind of savages nor that of primitive or archaic humanity, but rather mind in its untamed state as distinct from mind cultivated or domesticated for the purpose of yielding a return." [Trans.]

are not the outcome of function-producing "applications" rather than of "groupings" in the operational sense (see Section 12 above).

Lévi-Strauss' "natural logic" is the counterfoil to the "pre-logical" mentality postulated by Levy-Brühl, whose positivism he rejects in principle, as we would. But it seems to us that just as Levy-Brühl had earlier gone too far in one direction, so he went too far in the opposite direction in his posthumous retractions. There is no "primitive mentality," but there may well be a "pre-logic" in the sense of a pre-operational level of thought or of a level to begin with limited to concrete operations (see Section 12). "Participation" is a notion full of interest if one sees in it, not some mystical tie despising contradiction and identity, but a relation, common with young children, which remains midway between the generic and the individual: a child of four or five will describe the shadow thrown on a table as taken from "the shade beneath the trees" or as the shade of night; he does not include shadows in some general class; but this does not mean that he spatially "transports" the shade under the trees to the table, though for lack of anything better he may say this is what he means; rather, there is a kind of immediate fusion of objects which will only later, when their law is understood, become dissociated and reunited into a class. Even if participation is viewed merely as a variety of "analogical thought" (see *The Savage Mind,* p. 263), it would still be of interest as a pre-logic in the double sense of being *anterior* to an explicit logic and of being *preparatory* for the latter.

Unquestionably, the kinship systems described by Lévi-Strauss bear witness to a much more advanced logic. But these, as need hardly be mentioned and is perfectly obvious to the ethnographer, are not the products of individual invention (Tyler's "primitive philosopher"); it is long-term collective elaboration that has made them possible. They depend, therefore, as do linguistic structures, whose power likewise surpasses the resources of individuals, on *institutions*.[10] If the concept of self-regulation or equilibrium has any sense at all, the logic or pre-logic of the members of a given society cannot be adequately gauged by already crystallized cultural products; the real problem is to make out how the ensemble of these collective instruments is utilized in the everyday reasoning of each individual. It may very well be that these instruments are of a level visibly superior to that of western logic—Lévi-Strauss reminds us that there are plenty of natives who can "calculate" the implicit relations of a kinship system exactly.[11] But the kinship systems are finished systems, already regulated, and of limited scope. What we want to know about is individual inventions.

So we would hold that the question remains open so long as no careful studies of the operational level of both adults and children in a variety of societies have been carried out in systematic fashion. Such researches are difficult because they presuppose a good training in psychology—in particular, in the technique of operational examination (which calls for free conversation, not

[10] By way of analogy, the way in which termites construct their nest does not give unequivocal information about their geometrical behavior in other situations.

[11] See *The Savage Mind*, p. 251: the case of the Ambrym native described by Deacon.

the standardized questioning of "tests" to which most psychologists are accustomed)—sufficient anthropological knowledge, and mastery of the language of the subjects. We know of only a few such studies. One of these concerns the famous Arunda of Australia and seems to indicate a systematic retardation in the formation of concepts of conservation, though these are eventually acquired; in this particular case there seems to be access to the first layers of the level of concrete operations (see Section 17). But the Arundas' propositional operations (combinatory, and so on) remain to be examined; and, above all, many other societies must be studied from this kind of perspective.

As for the functional aspects of social structures, as soon as one admits partial auto-construction it seems impossible to abstract from them. Granting that factors of utility alone cannot account for the formation of a structure, they do suggest what some of the problems were to which this formation furnished a response. Taking them into account leads, therefore, to a rapprochement between "formation" and "response," as in Waddington's theories (see Section 10). It should also be mentioned that it happens frequently that a structure changes its function to meet new social needs.

None of the preceding observations throw doubt upon the positive, that is, specifically structuralist, aspects of Lévi-Strauss' analyses. Their sole purport is to induce this structuralism to leave its splendid isolation. Because it installs itself from the start in finished products, the traits which are perhaps most characteristic of human activity, even in its cognitive aspect, tend to be overlooked. Whereas other animals cannot alter themselves except by changing their species, man can transform himself by transforming the world and can structure himself by constructing structures; and these structures are his own, for they are not eternally predestined either from within or from without. So, then, the history of intelligence is not simply an "inventory of elements"; it is a bundle of transformations, not to be confused with the transformations of culture or those of symbolic activity, but antedating and giving rise to both of these. Granting that reason does not evolve without reason,[12] that it develops by virtue of internal necessities which impose themselves in the course of its interactions with the external environment, nevertheless reason has evolved, from the level of the animal or the infant to the structural anthropology of Lévi-Strauss himself.

[12] See *The Savage Mind*, p. 252: "Language, an unreflecting totalization, is human reason, which has its reason and of which man knows nothing." [Trans.]

VII

STRUCTURALISM AND PHILOSOPHY

20. *Structuralism and Dialectic*

In this chapter we shall take up just two general questions raised by structuralist investigation. The list of questions could be indefinitely extended, for now that the fashion has seized hold there is hardly any contemporary philosopher who does not go along with it, and the novelty of the fashion obscures the oldness of the method as employed in the sciences, since science is easily ignored in certain types of philosophy.

The first of the two problems we shall discuss is inescapable. To the extent that one opts for structure and devaluates genesis, history, and function or even the very activity of the *subject* itself, one cannot but come into conflict with the central tenets of dialectical modes of thought. It is therefore not surprising, and it is extremely instructive, to find Lévi-Strauss devoting almost the entire concluding chapter of *La Pensée sauvage* to a discussion of Sartre's *Critique de la raison dialectique*. An examination of this debate seems to us all the more in order because both of the antagonists appear to us to have forgotten the fundamental fact that in the domain of the sciences themselves structuralism has always been linked with a constructivism from which the epithet "dialectical" can hardly be withheld—the emphasis upon historical development, opposition between contraries, and *"Aufhebungen"* (*"dépassements"*) is surely just as characteristic of constructivism as of dialectic, and that the idea of wholeness figures centrally in structuralist as in dialectical modes of thought is obvious.

The principal components of dialectical thought as we find it in Sartre are constructivism and its corollary, historicism. We earlier touched on Lévi-Strauss' general critique of theories which assign a privileged status to history; Sartre is there singled out for special mention. The difficulties attaching to his view of the *I* and his notion of the *We* as no more than an *I* raised to the second power, hermetically sealed off from other *We*'s, are also pointed up. Though this last point is well taken, it should be mentioned that Sartre's subjectivist difficulties are the remains of his earlier existentialist phase; it is because his dialectic has not been schooled in the sciences but is merely doctrinal that it has not succeeded in erasing these vestiges of existentialism, for the dialectic of scientific thought implies, precisely, a reciprocity between perspectives. Sartre's constructivism we would defend, despite Lévi-Strauss' objections, except that we would deny what Sartre affirms, namely, that constructivism is peculiarly philosophical and alien to science. Sartre's depiction of science is almost entirely derived from positivism and its "analytic" method. Now not only is positivism, a movement in *philosophy,* not the same as *science* (of which it gives a systematically distorted picture), but —as Meyerson often pointed out—even the most positivistic scientists do not act on the credo they expound in their prefaces; they do just about the opposite of what dogma requires as soon as they turn to the analysis and explanation of experience. It is one thing to accuse them of insufficient self-knowledge or epistemological sophistication, but quite another simply to assimilate their scientific work to positivism.

But this means that Lévi-Strauss' conception of the connection between dialectical reason and scientific thought, though more adequate than Sartre's, is also open to objection: it is alarmingly modest as to the requirements of science and obliges us to grant a much more important role to dialectical processes than Lévi-Strauss himself seems to want. Not that there is an inherent conflict between structuralism and dialectic; rather, Lévi-Strauss' version has been relatively static and ahistorical, and this is what has led him to underestimate the importance of dialectical processes.

What, for Lévi-Strauss, is dialectical reason? If we understand him aright, it is always "constitutive" (*The Savage Mind,* p. 246) in the sense of being venturesome, building bridges and crossing them, whereas analytic reason separates because it wants not only to understand but to control. "Dialectical reason," Lévi-Strauss tells us, "is not . . . something *other than* analytic reason . . . it is *something additional in* analytic reason" (*The Savage*

Mind, p. 246); it is analytic reason's own effort to transcend itself. But are we forcing the words if we say this comes down to a complementarity according to which synthetic reason's inventiveness and progressiveness make up for the lack of these in analytic reason while the job of verification remains reserved for the latter? The distinction is, of course, essential and, equally of course, there are not two reasons but two attitudes or two "methods" (in the Cartesian sense) which reason may adopt. Still, to describe the work of construction for which the dialectical attitude calls simply as a matter of "throwing out bridges over the abyss of a human ignorance whose further shore is constantly receding" (*The Savage Mind,* p. 246) is insufficient. It is often construction itself which begets the negations along with the affirmations, and the syntheses (*dépassements*) whereby they are rendered coherent as well.

This Hegelian or Kantian pattern is not a merely conceptual or abstract pattern such as would be of no interest to either the sciences or structuralism. It corresponds to a progression which is inevitable once thought turns away from false absolutes. In the realm of structure it matches a recurrent historical process well described by G. Bachelard in one of his best books, *La Philosophie du non.* Its principle is that, given a completed structure, one negates one of its seemingly essential or at least necessary attributes. Classical algebra, for example, was commutative, but since Hamilton we have a variety of non-commutative algebras; Euclidean geometry has by "negation" (of the parallel postulate) engendered the non-Euclidean geometries; two-valued logic with its principle of excluded middle has, through Brouwer's denial of the unrestricted validity of this principle (in particular, its validity in reasoning about Cantorian sets), become supplemented by multi-valued logics, and so on. In logic and mathematics, construction by negation has practically become a standard method; given a certain structure, one tries, by systematic negation of one after another attribute, to construct its complementary structures, in order later to subsume the original together with its complements in a more complex total structure. Griss' "negationless logic" goes so far as to "negate" negation. Furthermore, when what is in question is to determine whether it is system *A* which presupposes *B* or *B* which presupposes *A* (for example, whether ordinals or cardinals are prior, concepts or judgments, and so on) we can be quite sure that dialectical circles or interactions will always in the end replace linear orders of prior and posterior.

In physics and biology there is something analogous to what we called "construction by negation," though here it derives from what Kant called "real opposition."[1] Need we remind the reader of the oscillations back and forth between a corpuscular and a wave theory of light, or the reciprocities between electrical and magnetic processes of which we know since Maxwell? Here, as in the domain of abstract structures, the dialectical attitude seems essential to the full working out of structures; dialectic is both complementary to and inseparable from analytic, even formalizing, reason; so the "something more" which

---

[1] See L. Apostel's interesting chapter on logic and dialectic in *Logique et connaissance scientifique* (Pléiade), where this Kantian notion of a contrast between real and logical opposition is discussed at length.

Lévi-Strauss grudgingly allows to it is not just the courage to "throw out bridges": dialectic over and over again substitute "spirals" for the linear or "tree" models with which we start, and these famous spirals or non-vicious circles are very much like the genetic circles or interactions characteristic of growth.

This brings us back to the problem of history and Althusser's and Godelier's attempts to subject Marx's work, despite the essential role it assigns to historical development in its sociological interpretations, to structuralist analysis. That there is a structuralist strand in Marx, something just about halfway between what we called "global" and "analytic" structuralism, is obvious, since he distinguishes "real infrastructures" from "ideological superstructures" and describes the former in terms which, though remaining qualitative, are sufficiently precise to bring us close to directly observable relations. Althusser, who means to furnish Marxism with an epistemology, tries therefore, and with full justification, to differentiate the Marxist from the Hegelian dialectic and to reformulate the former in modern structuralist terms.

According to Althusser,[2] the "Hegelianism" of the young Marx is quite debatable; Marx took off rather from problems set by Kant and Fichte. Whether Althusser is right on this point we cannot judge. It is a corollary of two much more fundamental observations. The first is that for Marxism, in contrast to idealism, to *think* is to *produce*, thought being a kind of "theoretical practice" which is not so much the work of an individual subject as the outcome of interactions between the subject and his

[2] Althusser, *Pour Marx* (Paris: Maspero, 1965). [Trans.]

personal environment, into which social and historical factors enter as well; it is in this light that Althusser interprets Marx's famous passage where "the totality of the real" as a *Gedankenconcretum* is said to be "in reality a product of thought and conception."[3]

We also accept Althusser's second observation, namely, that dialectical contradiction in Marx bears no resemblance to the Hegelian, which is, in the final analysis, reducible to an identity of contraries, whereas for Marx dialectical contradiction is the result of "overdetermination" (*surdétermination*), that is, if we understand him right, a necessary consequence of the inseparability of interactions. Similarly, Althusser rightly points up the difference between the Hegelian and the Marxist notions of "totality."

It is this notion of "overdetermination"—the sociological counterpart to certain forms of causality in physics —which prompts Althusser to insert the contradictions inherent in the relations of production or the contradictions between these and the forces of production, in short, all the apparatus of Marxist economics, into a transformational system whose structure and principles of formalization he tries to articulate. Althusser has been chided for his formalism, but this is the current and unfounded criticism of all serious structuralist theories. The chief objection urged against him is that—at least in the eyes of some critics—he has too low an estimate of things human; but if the values of the "person" (often regrettably confused with those of the ego) are taken to be less im-

---

[3] Althusser uses the passage as an epigraph, *Pour Marx*, p. 186. [Trans.]

portant than the constructive activities of the epistemic subject, the characterization of knowledge as production is in agreement with one of the best established traditions of classical Marxism.

Godelier, in a footnote to his article "Système, structure, et contradiction dans le Capital,"[4] indicates, with great lucidity, how much work remains to be done on the relations between historic structures and their transformations. Social structures are comparable to mathematical "categories" (in the sense explained in Section 6 —sets of objects and their possible mutual "applications"). It is not at all difficult to determine which functions are compatible and which incompatible with a given social structure. The hard question is, given a systematic ensemble of such structures, how do the modalities of their mutual connections "induce a *dominant* function within one of the structures so connected"? Not until contemporary structural analysis has perfected its methods by studying historical and genetic transformations will it be able to furnish the answer. Though Godelier (whose rounding off of Althusser's analysis of contradiction in Marx is quite remarkable) stresses the "priority of the study of structures to that of their genesis or evolution," and notes that Marx followed this procedure himself in opening *Das Kapital* with a theory of value, he can nevertheless be said to approach the question from this perspective: Let us not forget that, even in the domain of psychogenesis, genesis is never anything except (see Sections 12 and 13) the transition from one structure to another, and while the second structure is

[4] *Les Temps modernes* (1966), p. 857, Note 55.

explained in terms of this transition, the transition itself can only be understood in transformational terms if both of its termini are known. Godelier's final conclusion is worth citing in full, because it summarizes not only our objections to Lévi-Strauss but also the leading ideas of this work as a whole.

> Anthropology could no longer challenge history, nor history anthropology; the opposition between psychology and sociology, sociology and history, would become sterile. For the possibility of a "science" of man would, in the final analysis, depend upon the possibility of discovering the laws governing the operation, evolution, and internal relations of social structures . . . the method of structural analysis will, in other words, have to be generalized so as to become capable of explaining the conditions of variation and evolution of structures and their functions.[5]

For a structuralism of this sort, structure and function, genesis and history, individual subject and society are—once the instruments of analysis have been refined—inseparable, the more so the more it perfects its analytic tools.

21. *Structuralism without Structures*

At the opposite end there is Michel Foucault's *Les mots et les choses*. Written in a dazzling style, full of unexpected and brilliant ideas, tremendously erudite, it keeps only the negative aspects of contemporary structuralism.

[5] *Les Temps modernes*, p. 864.

This "archaeology of the human sciences" (as the work is subtitled) seems, in the end to be nothing but a search for conceptual archetypes, chiefly tied to language. Foucault has it in for man; the human sciences he views as a merely momentary outcome of "mutations," "historical a priorities," "epistemes"; these follow one another in time, but their sequence has no rationale. Not until the nineteenth century did man become the object of scientific study, and the human sciences will perish as surely as they came into existence. We do not and cannot know what new variety of *episteme* will take their place.

Curiously, Foucault locates one of the reasons for their imminent extinction in structuralism itself. Structuralism allows "for the possibility, even sets itself the task, of purifying the old empirical reason by constructing formal languages; it wants to carry out a second critique of pure reason, which takes its departure from new forms of the mathematical *a priori*" (p. 394). By generalizing the powers of language in this way, "by stretching its possibilities to the breaking point, [structuralism] spells the end of man. In reaching the summit of all possible speech, man does not attain to its heart but to the boundary of what limits it: death roams about in this region, thought becomes extinguished, the original promise indefinitely remote" (pp. 294ff.). And yet, "structuralism is not a new method; it is the roused and uneasy conscience of modern science" (p. 221).

Skeptical epistemologies have a real function, that of raising new problems by undermining easy solutions. What we would want of Foucault is, accordingly, that he prepare the way for a second Kant to reawaken us, along with himself, from dogmatic slumber. In particular, we would expect the author of a work with such revolutionary intentions to offer a constructive critique of the human sciences, an intelligible account of the new-fangled notion of *episteme,* and an argument that would justify his restrictive conception of structuralism. But we are disappointed on all three counts. Beneath the cleverness there are only bare affirmations and omissions; it is up to the reader to make connections and to construct arguments as best he can. For example: the human sciences are not just "false sciences," they are not sciences at all, for, says Foucault, the very configuration by which their "positivity" is defined and which gives them roots in modern *episteme* consigns them to a place outside the sciences. If one inquires: "why, then, are they *called* sciences?" Foucault repeats his archaeological definition, according to which all inquiry dons the name and adopts the models of science (p. 378), and seems to think this a sufficient answer. No demonstration of these unheard-of assertions is forthcoming. All that is said is: (1) the "configuration by which their positivity is defined" is a "trihedron" (invented by Foucault, pp. 355–359) whose three faces are: (a) mathematics and physics (b) biology, economics and linguistics (which—see p. 364—are not human sciences) (c) philosophical reflection; (2) the human sciences do not appear on any of the three faces; *therefore,* they are not sciences. (3) Foucault's "archaeological definition of their roots" easily answers the question whence the delusion that the human sciences *are* genuine sciences, since all that these "definitions" amount to is a retrospective account of what *did* happen as if it were deductible *a priori* from an acquaintance with their *episteme* ("history shows that all that has been thought will be rethought in a thought that has not as yet seen the light of day"—p. 383).

Instead of criticizing the human sciences in terms which its practitioners would accept, Foucault redefines "human science"; this makes his task rather too easy. For example, as already mentioned, linguistics is not a human science except when it deals with the way in which individuals or groups represent words to themselves (p. 369). Scientific psychology is the creature of the "new norms which industrial society has since the nineteenth century imposed on individuals" (it would be interesting to know which these are).[6] It is resolutely severed from its roots in biology. What remains is: analysis of individual representations (though no psychologist could possibly be satisfied with this) and, as is to be expected, the Freudian unconscious, which Foucault appreciates all the more because it prefigures the end of man, in the sense that it dissolves the privileged object status of consciousness. What Foucault forgets is that the whole of cognitive life is linked to structures which are just as unconscious as the Freudian Id, but which reconnect knowledge with life in general.

None of this would be very important if Foucault's critique, however biased, rested upon a real discovery. His concept of *episteme* looks promising at first; it seems to call for some sort of epistemological structuralism, which would be very welcome. Foucault's *epistemes* do not form a Kantian system of *a priori* categories, for, un-

---

[6] Foucault neglects to mention Helmholtz, Hering, and many other victims of the "new-fangled norms of industrial society," Darwin himself (one of the founders of scientific psychology) included.

like Lévi-Strauss' human reason, they are neither necessary nor permanent; they simply follow one another in
the course of history. They are not systems of observable
relations resulting from intellectual habits, nor are they
constraints upon thought in general which somehow become manifest at a given moment in the history of science. They are "historical a priorities," like Kant's
transcendental forms "conditions for knowledge" but,
unlike these, conditions which apply for only a limited
period and which, when their vein has been exhausted,
yield to others.

Foucault's *epistemes* are strikingly reminiscent of Th.
S. Kuhn's "paradigms,"[7] and at first sight Foucault's
analysis, because of its structuralist ambitions, even seems
more profound than Kuhn's. Foucault's program, were he
able to carry it through, would lead to the discovery of
strictly epistemological structures that would show how
the fundamental principles of the science of a given
period are connected with one another, whereas Kuhn
merely describes them and analyzes the intellectual crises
which resulted in "mutations." Yet Foucault's more am-
bitious program requires a method, and this is where he
falls short, for instead of inquiring under what conditions
one may speak of the reign of a new *episteme* and what
are the criteria by which to judge the validity or invalidity
of alternative interpretations of the history of science,
he relies on intuition and substitutes speculative improvi-
sation for methodical procedure.

He has no canon for the selection of an *episteme's*

[7] See Thomas S. Kuhn, *The Structure of Scientific Revolutions*
(Chicago: University of Chicago Press, 1962).

characteristics; important ones are omitted, and the choice between alternative ones is arbitrary. Furthermore, heterogeneous attributes which happen to be found together in the same historical period, though they really belong to different levels of thought, are treated as of one kidney. Thus Foucault's characterization of contemporary *episteme*, the "trihedron" we mentioned earlier, is arbitrary from every point of view. He sets up an idiosyncratic classification of the human sciences which excludes linguistics and economics (except when they touch, not on man, but on individuals or finite groups) and which turns psychology and sociology into vagrants because they cannot settle down on any one of the three facets of his trihedron. Clearly, this *episteme* is a creation of Foucault's, not a transcription of contemporary scientific trends. Furthermore, his trihedron is static, while the fundamental trait of the sciences today is the multiplicity of their interactions, which tend to form a system closed upon itself with many cross-linkings: thermodynamics with information theory, psychology with ethology and biology, psycholinguistics with generative grammar, logic with psychogenetics, and so on. Last, Foucault reserves a distinct facet of his trihedron for philosophical reflection, whereas epistemology is in fact increasingly internal to the several sciences, dependent upon their cyclic arrangement, subject to shifts as interdisciplinary relations become modified (all this, by the way, goes to confirm Foucault's oft-repeated assertion [as on p. 329] that man, that "strange double being," is "empirico-transcendental.")

The table on page 87, which reduces seventeenth- and eighteenth-century *episteme* to linear structures and taxonomic trees, is a particularly clear example of the dangers of "homogenization." Biology did indeed remain arrested at the taxonomic level (which, structurally considered, amounts to an exceedingly elementary "group" subject to numerous restrictive conditions and not really capable of expansion except by juxtaposition), but surely seventeenth- and eighteenth-century mathematics and physics went far beyond this (infinitesimal analysis, Newton's law of action and reaction and so on). To maintain that they together constitute one *episteme* simply because they are synchronous is to be too easy a prey for history, and this despite the fact that Foucault's intellectual archaeology is meant to set us free from history!

Foucault completely disregards this fundamental problem of levels because it does not fit in with his personal "archaeological" *episteme*. But he pays an exorbitantly high price for this; the sequence of *epistemes* is thereby, and deliberately, rendered incomprehensible. Foucault seems to relish this incomprehensibility. His *epistemes* follow upon, but not from, one another, whether formally or dialectically. One *episteme* is not affiliated with another, either genetically or historically. The message of this "archaeology" of reason is, in short, that reason's self-transformations have no reason and that its structures appear and disappear by fortuitous mutations and as a result of momentary upsurges. The history of reason is, in other words, much like the history of species as biologists conceived of it before cybernetic structuralism came on the scene.

To call Foucault's structuralism a structuralism without structures is, accordingly, no exaggeration. All the negative aspects of static structuralism are retained—the devaluation of history and genesis, the contempt for functional considerations; and, since man is about to disappear, Foucault's ouster of the subject is more radical than any hitherto. Indeed, his structures are in the end mere diagrams, not transformational systems. In this irrationalism only one thing is fixed, language itself, conceived as dominating man because beyond individuals. Yet the being of language is deliberately kept mysterious; only its "enigmatic insistency" is fondly stressed.

But none of this takes away from the fact that Foucault's corrosive intelligence has performed a work of inestimable value: that of demonstrating that there cannot be a coherent structuralism apart from constructivism.

CONCLUSION

BEFORE WE TURN to a summary of the theses which this little book has tried to separate out from the principal structuralist positions, we should remind the reader that, though a good many applications of the method are new, structuralism itself is not. It has a long history, which forms part of the history of the sciences, even if in comparison with the hypothetico-deductive method it is of comparatively recent origin.

That it should have taken so long before its possibility was discovered is, of course, primarily due to the natural tendency of the human mind to proceed from the simple to the complex, hence, to neglect interdependencies and systematic wholes until such time as problems of analysis force them upon our attention. It is also due to the fact that structures are not observable as such, being located at levels which can be reached only by abstracting forms of forms or systems of the nth degree; that is, the detection of structure calls for a special effort of reflective abstraction.

Nevertheless, structuralism has a comparatively long history, and one lesson to be drawn from this is that structuralism cannot be a particular doctrine or philosophy; had it been that, it would long have been left behind. Structuralism is essentially a method, with all that this term implies—it is technical, involves certain intellectual obligations of honesty, views progress in terms of gradual approximation. Calling structuralism's history to mind is salutary for yet another reason: no matter how open-minded the properly scientific attitude, one can only be disturbed by the current modishness of structuralism, which weakens and distorts it. Only by withdrawing somewhat from the immediate present will we enable authentic (analytic) structuralism to appraise what is currently being said in its name.

This much recalled, the most important conclusion to be distilled from our series of investigations is that the study of structure cannot be exclusive and that it does not suppress, especially in the human sciences and in biology, other dimensions of investigation. Quite the contrary, it tends to integrate them, and does so in the way in which all integration in scientific thought comes about, by making for reciprocity and interaction. Whenever we detected a certain exclusiveness in a particular structuralist position, the next or some earlier chapter showed us that the models appealed to in the effort to justify this limitation or "hardening" were evolving in just the opposite direction from that supposed. Thus, to recall just one example, after linguistics had earlier inspired all sorts of fruitful but somewhat one-sided ideas, there came the unexpected inversions of Chomsky to broaden these overly narrow views.

Our second general conclusion is that the search for structures cannot but result in interdisciplinary coordinations. The reason for this is quite simple: if one tries to deal with structures within an artificially circumscribed domain—and any given science is just that—one very soon hits on the problem of being unable to locate the entities one is studying, since structure is so defined that it cannot coincide with any system of observable relations, the only ones that are clearly made out in any of the existing sciences. For example: Lévi-Strauss assigns his structures to a system of conceptual schemes somewhere midway between "infrastructures" and conscious systems of conduct or ideology, because "ethnology is first of all a psychology." And he is quite right, for, as psychogenetic studies have shown, the mechanisms on which the individual subject's acts of intelligence depend are not in any way contained by his consciousness, yet they cannot be explained except in terms of "structures" (that is, only by appealing to the very structures we discussed in Chapter II—"groups," "networks," "semigroups," and so on—can we make sense of the *intelligence* of intelligent behavior). But we, if asked to "locate" these structures, would carry Lévi-Strauss' suggestion one step further; we would assign them a place somewhere midway between the nervous system and conscious behavior because, to adapt his locution, "psychology is first of all a biology." One might even want to take the next step, except that, since the sciences form a cycle rather than a linear series, such a descent from biology to physics would only be preparation for a subsequent return to mathematics, which in the end would bring us back to —well, what exactly? Let us say, to man himself, so as not to force the option between the human organism and the human mind.

To return now to our conclusions. One thing seems evident from the foregoing comparative study—"structures" have not been the death of the *subject* or its activities. True, there is much here that stands in need of clarification, and some philosophical traditions have piled up such confusion on the topic that what *they* call the subject *is* undermined. Thus, in the first place, structuralism calls for a differentiation between the *individual subject*, who does not enter at all, and the *epistemic subject*, that cognitive nucleus which is common to all subjects at the same level. In the second place, the always fragmentary and frequently distorting grasp of consciousness must be set apart from the achievements of the subject; what he knows is the *outcome* of his intellectual activity, not its mechanisms. Now after such precipitation of the "me," the "lived," from the "I," there remains the subject's "operations," that which he "draws out" from the general coordinations of his acts by reflective abstraction. And it is these operations which constitute the elements of the structures he employs in his ongoing intellectual activity.

It might seem that the foregoing account makes the *subject* disappear to leave only the "impersonal and general," but this is to forget that on the plane of knowledge (as, perhaps, on that of moral and aesthetic values) the subject's activity calls for a continual "de-centering" without which he cannot become free from his spontaneous intellectual egocentricity. This "de-centering" makes the subject enter upon, not so much an already available and therefore external universality, as an uninterrupted process of coordinating and setting in reciprocal relations. It is the latter process which is the true "generator" of structures as constantly under construction and reconstruction. The subject exists because, to put it very briefly, the being of structures consists in their coming to be, that is, their being "under construction."

The justification for this last assertion is furnished by the following conclusion, like all the preceding drawn from the comparative study of distinct domains of science: *There is no structure apart from construction,* either abstract or genetic. Now as we have seen, these two kinds of construction are not as far removed from one another as is commonly supposed. Since Gödel, logicians and students of the foundations of mathematics distinguish between "stronger" and "weaker" structures, the stronger ones not being capable of elaboration until after the construction of the more elementary, that is, "weaker" systems yet, conversely, themselves necessary to the "completion" of the weaker ones. The idea of a formal system of abstract structures is thereby transformed into that of the construction of a never completed whole, the limits of formalization constituting the grounds for incompleteness, or, as we put it earlier, incompleteness being a necessary consequence of the fact that there is no "terminal" or "absolute" form because any content is form relative to some inferior content and any form the content for some higher form. If Gödel's theorem may fairly be interpreted in this way, "abstract construction" is merely the formalized inverse of "genesis," for genesis too proceeds by way of reflective abstraction, though it starts lower down on the scale. It is only natural that, in areas where the genetic data are unknown and beyond recovery, as in ethnology, one puts a good face on a bad situation and pretends that genesis is quite irrelevant. But in areas where genesis obtrudes on everyday obser-

7

vation, as in the psychology of intelligence, one cannot help but become aware that structure and genesis are necessarily interdependent. Genesis is simply transition from one structure to another, nothing more; but this transition always leads from a "weaker" to a "stronger" structure; it is a "formative" transition. Structure is simply a system of transformations, but its roots are operational; it depends, therefore, on a prior formation of the instruments of transformation—transformation rules or laws.

The problem of genesis is not just a question of psychology; its framing and its solution determine the very meaning of the idea of structure. The basic epistemological alternatives are predestination or some sort of constructivism. For the mathematician it is, of course, tempting to believe in Ideas and to think of negative or imaginary numbers as lying in God's lap from all eternity. But God himself has, since Gödel's theorem, ceased to be motionless. He is the living God, more so than heretofore, because he is unceasingly constructing ever "stronger" systems. Passing from "abstract" to "real" or "natural" structures, the problem of genesis becomes all the more acute. Only if we forget about biology can we be satisfied with Chomsky's theory of the innateness of human reason or with Lévi-Strauss' thesis of the permanence of the human intellect. The epistemological alternatives we mentioned apply even in the realm of organic structures, which may be viewed either as the products of an evolving process of construction or as predestined from the beginning and for all time by the original DNA molecule. In short, the same problem turns up no matter where we look. Suffice it to note, by way of conclusion, that genetic construction *is* being studied, that the structuralist perspective has, if anything, emphasized the importance of such investigation, and that, as a result, some such synthesis as we saw taking shape in linguistics and in the psychology of intelligence is now establishing itself elsewhere too.

What about functionalism? Since, as we argued, structuralism does not by any means eliminate the epistemic subject, and since its structures cannot be understood apart from their genesis, the concept of function has obviously lost none of its value; all talk about self-regulation involves the idea of function. Here again *de jure* arguments serve to corroborate arguments *de facto*. In the realm of natural structures, the denial of activity leads to the postulation of an entity—the subject, society, life, or what have you—which might serve as "structure of all structures" since (unless, with Foucault, one assumes a sequence of separate and contingent *epistemes*) structures can live only in systems. Now, as we have come to see more clearly through Gödel but knew long before, the ideal of a structure of all structures is unrealizable. The *subject* cannot, therefore, be the *a priori* underpinning of a finished posterior structure; rather, it is a center of activity. And whether we substitute "society" or "mankind" or "life" or even "cosmos" for "subject," the argument remains the same.

So, to repeat, structuralism is a method, not a doctrine, whose doctrinal consequences have been quite various. Because it is a method, its applicability is limited; that is, if, precisely on account of its fruitfulness, it has become connected with other methods, it admits the legitimacy of these other methods. Far from ousting genetic or functionalist studies, it rather implements them by giving them the benefit of its very powerful instruments of analysis, particularly in those border areas where new connections must be made. On the other hand, its methodological character also makes for a certain openness. Structuralism is as willing to get as to give; only, being a recent arrival and still full of unexpected riches, the exchange between it and more established methods has been somewhat uneven.

Just as the structuralism of the Bourbaki has already expanded into a movement calling for more dynamic structures (the categories with their functional emphasis), so the other current forms of structuralism are no doubt big with future developments. And since an immanent dialectic is here at work, we can be sure that the denials, devaluations, and restrictions with which certain structuralists today meet positions which they regard as incompatible with their own will one day be recognized to mark those crucial points where new syntheses overtake antitheses.

SELECTED BIBLIOGRAPHY

Bourbaki, Nicholas. "L'architecture des mathématiques," in F. Le Lionnais, *Les grands courants de la pensée mathématique*. Paris, 1948.

Chomsky, Noam. *Syntactic Structures*. The Hague: Mouton, 1957.

Casanova, G. *L'algèbre de Boole*. Paris: Presses Universitaires de France, collection "Que sais-je?" No. 1246, 1967.

Foucault, Michel. *Les mots et les choses*. Paris: Gallimard, 1966.

Lacan, J. *Ecrits*. Paris: Editions du Seuil, 1966.

Lévi-Strauss, Claude. *Elementary Structures of Kinship*. Edited by Rodney Needham Bell. Translated by John R. Von Sturmer. Boston: Beacon, 1969.

———. *The Savage Mind*. Chicago: University of Chicago Press, 1966.

———. *Structural Anthropology*. New York: Basic Books, 1963.

Lewin, Kurt. *Field Theory in Social Science*. Edited by Dorwin Cartwright. New York: Harper, 1951.

Parsons, Talcott. *Structure and Process in Modern Societies*. Glencoe, Ill.: The Free Press, 1960.

Piaget, Jean. *Traité de logique*. Paris: Colin, 1949.

———. *Biologie et connaissance*. Paris: Gallimard, 1967.

——— et al., *Logique et connaissance scientifique. Encyclopédie de la Pléiade*, vol. XXII.

de Saussure, Ferdinand. *Course in General Linguistics*. Edited

by C. Bally and A. Séchehaye. Translated by Wade
Baskin. New York: Philosophical Library, 1959.

De Zwaart, H. Sinclair. *Acquisition du langage et développement de la pensée*. Paris: Dunod, 1967.

Tinbegren, J. "De quelques problèmes posés par le concept de structure." *Revue d'Economie politique*, LXII, 27–46.

INDEX